MY SISTERS TELEGRAPHIC

My Sisters Telegraphic

Women in the Telegraph Office,
1846–1950

THOMAS C. JEPSEN

OHIO UNIVERSITY PRESS • *Athens*

Ohio University Press, Athens, Ohio 45701
© 2000 by Thomas C. Jepsen
Printed in the United States of America
All rights reserved

Ohio University Press books are printed on acid-free paper ⊗ ™

09 08 07 06 05 04 03 02 01 00   5 4 3 2 1

Library of Congress Cataloging-in-Publication Data

Jepsen, Thomas C.
    My sisters telegraphic : women in the telegraph office, 1846–1950 / Thomas C.
    Jepsen.
      p. cm.
    Includes bibliographical references and index.
    ISBN 0-8214-1343-0 — 0-8214-1344-9 (pbk.)
      1. Telegraphers—United States—History. I. Title.
    HD6073.T252 U65 2000
    331.4'813841'0973—dc21
    00-040654

Contents

Illustrations

Tables

Acknowledgments

I would like to thank all those who thoughtfully responded to my written requests for information, answered my telephoned and e-mailed queries, or otherwise participated in the "telegraphic network" that furthered my research and sent me off in new and fruitful directions: Melodie Andrews, Mankato State University, Mankato, Minnesota; Christopher T. Baer, Hagley Museum and Library, Wilmington, Delaware; Elizabeth Bailey, State Historical Society of Missouri, Columbia, Missouri; Kathryn L. Barton, Guernsey Memorial Library, Norwich, New York; Evelyn Belknap, Daughters of Utah Pioneers, Salt Lake City, Utah; Lynne Belluscio, Leroy Historical Society, Leroy, New York; Kathryn L. Bridges, Charles A. Cannon Memorial Library, Concord, North Carolina; Shirley Burman, Curator, "Women and the American Railroad," Sacramento, California; Elwood W. Christ, Adams County Historical Society, Gettysburg, Pennsylvania; Paul Cyr, New Bedford Free Public Library, New Bedford, Massachusetts; Merle Davis, State Historical Society of Iowa City, Iowa; Amy Doliver, Office of Chenango County Historian, Norwich, New York; Melissa L. Dunlap, Niagara County Historical Society, Lockport, New York; Charles L. Eater, Mifflin County Historical Society, Lewistown, Pennsylvania; Marie Edgar, Mine au Breton Historical Society, Potosi, Missouri; Edwin Gabler, University of Maryland, College Park; Ruth B. Gembe, Alexander Hamilton Memorial Free Library, Waynesboro, Pennsylvania; Ellen Hagney, Lowell Historic Preservation Society, Lowell, Massachusetts; Robert S. Harding, Archives Center, National Museum of American History, Washington, D.C.; Alice Henson, genealogical researcher, Jefferson City, Missouri; Elaine Pollock Lundquist, telegrapher, La Mesa, California; Karen W. Mahar, Siena College, Londonville, New York; Martha Mayo,

University of Lowell Special Collections, Lowell, Massachusetts; Genevieve G. McBride, University of Wisconsin–Milwaukee; Carolyn J. Norris, Family History Research, Springfield, Illinois; Coralee Paull, genealogical researcher, St. Louis, Missouri; Robin Rummel, Johnstown Heritage Museum, Johnstown, Pennsylvania; William F. Strobridge, Wells Fargo Bank, San Francisco, California; Marion Strode, Chester County Historical Society, West Chester, Pennsylvania; Jan Taylor, Grout Museum, Waterloo, Iowa; and Mary K. Witkowski, Bridgeport Public Library, Bridgeport, Connecticut. Thanks also to Gillian Berchowitz, Senior Editor, Ohio University Press, for her help in bringing this project to fruition. And a special thanks to my wife, Marsha, not only for her help and support but also for getting me an interview with her aunt, former telegrapher Virginia Brom Turner of Oskaloosa, Iowa.

MY SISTERS TELEGRAPHIC

Women in the Telegraph Industry

PICTURES of women working as telegraphers in nineteenth-century popular literature always seem to lack context. These women do not fit neatly into either of the predominant stereotypes of nineteenth-century women: the devoted wife and mother, secure in her domestic sphere, or the exploited factory operative, forced to work long hours to earn a subsistence living. Part of the difficulty in creating context is the absence of a cultural memory; we have largely forgotten that women ever did this sort of work. A hundred years ago, however, women telegraph operators were commonplace. Frances Willard, writing in 1897, noted that the sight of "a young woman presiding over the telegraph in offices and railway stations" was so ordinary "that one has ceased to have even a feeling of surprise at seeing them there."[1]

When the presence of women in the telegraph office did elicit comment, it was generally to note their exceptionalism. A writer for *Electrical World* observed in 1886 that a rail traveler stopping at a remote station in the deserts of the West was likely to see "a bright, neatly dressed, white-aproned young woman come to the door and stand gazing out at the train and watching the passengers with a half-pleased, half-sorry air." She is presumably pleased at the safe arrival of the train and its passengers, owing partly to her technical skills, and sorry that she will soon be left again to her solitude; for the writer notes, "This is the local telegraph operator, who has taken up her lonely life out here on the alkali desert amid the sage brush, and whose only glimpse of the world she has left behind her is the brief acquaintance with the trains which pass and repass two or three times during the day."[2]

In the mid-nineteenth century, women telegraph operators entered a challenging, competitive technological field in which they competed directly with men, demanding, and occasionally getting, equal pay and sometimes moving into management and senior technical positions. Women telegraphers constituted a subculture of technically educated workers whose skills, mobility, and independence set them apart from their contemporaries. The story of these women has remained untold, partly because the telegraph itself has been forgotten—and partly because these women were so far ahead of their time.

The role of the telegrapher in the mid-nineteenth century was similar to that of the contemporary software programmer/analyst. A rapidly growing industry had a sudden need for persons with technical skills, creating opportunities for ambitious women as well as men. To be a telegrapher, one had to be extremely literate and a good speller, be capable of learning Morse code, and have some knowledge of electricity and telegraphy.

Women played an important role in the telegraph industry from the 1840s onward, yet almost no written documentation on their activities exists. Numerous stories of male telegraphers who went from "rags to riches," like Thomas Edison and Andrew Carnegie, have entered the popular literature, but there are few corresponding stories about women.

This book describes daily life in the telegraph office and discusses the impact that women telegraphers had on culture and society, both in the United States and in other parts of the world. It discusses women's part in the telegraphers' labor movement and details the lives of some women telegraphers. Finally, it offers some thoughts on how women's presence in technical fields today has been influenced by the work of these pioneers.

The first chapter describes the chronological evolution of the telegraph industry in the United States and elsewhere and shows how the role of women telegraph operators changed over time. The next three chapters discuss daily life in the telegraph office, the relationship of female telegraphers to society, and women's issues in the telegraph office. The fifth chapter analyzes the literature that describes the work of the women telegraphers and their treatment in the cinema. The sixth chapter provides a chronological overview of women's involvement in the telegraphers' labor movement, and the final one offers some conclusions about the relationship of women telegraphers to women in technical professions today.

THE ENTRY OF WOMEN INTO THE TELEGRAPH INDUSTRY IN THE UNITED STATES

The introduction of the telegraph in the United States and Europe in the 1840s and the simultaneous development of the railroad system marked the beginning of the industrial age and a revolution in the speed of transportation and communication. Among the social consequences of this revolution were the development of a new middle class of clerks, managers, and office workers and a new category of workers who defined themselves in terms of the technological field in which they worked.

One of these new technical occupations—telegraphy—began as a relatively gender-neutral profession. Unlike many of the occupations women entered for the first time in the mid-nineteenth century, telegraphy admitted women to its ranks before its gender roles had solidified. During much of the nineteenth century, men and women performed the same tasks using the same equipment, working cooperatively and often anonymously at either end of the wire. Therefore, as the Canadian historian Shirley Tillotson points out in her essay, "We may all soon be 'first-class men,'" telegraphy provides a unique opportunity to study the relationship between gender and skill.[3]

The entry of women into the profession in the first ten years or so after the invention of the telegraph attracted little public notice. The appointment of Sarah G. Bagley as superintendent of the Lowell, Massachusetts, office of the New York and Boston Magnetic Telegraph Company on February 21, 1846, less than two years after Samuel Morse and Alfred Vail first publicly demonstrated their new invention, is generally regarded as a footnote to her earlier work as women's rights advocate and founder of the Lowell Female Labor Reform Association. To Bagley's contemporaries, her appointment was significant more for her class origins than her gender; the Lowell reform newspaper, the *Voice of Industry*, commended Paul R. George, manager of the telegraph company, on his "democracy" in choosing a member of the working class for the position: "This is what we call 'the people's' *democracy*, Miss Bagley having served ten years in the factories."[4]

As the telegraph lines began to spread across the United States in the late 1840s, the demand for telegraph operators quickly exceeded the supply, especially in rural areas. The entrepreneurs who organized the early telegraph companies quickly seized on the idea of employing women as operators, including members of their own families who often already had some hands-on experience with the new invention. John J. Speed, builder of the Erie and Michigan Telegraph line, wrote to Ezra Cornell, one of Morse's early associates, from Detroit, Michigan, in July 1849 to propose that women be hired as operators. Acknowledging that Cornell had warned him in the past to "avoid experiments," Speed gave notice that he was "making one more." He had hired a Mrs. Sheldon to run the telegraph office in Jackson, Michigan, and suggested that Cornell's sister Phoebe Wood could be hired to operate the Albion, Michigan, office. He cited both the scarcity of trained operators and shortage of funds as his reasons and expressed confidence in the ability of the women to perform the work: "Both are abundantly qualified to do the business better than any boy, or man, that we can afford to pay in those places."[5]

Phoebe Wood accepted the position as operator in Albion and shortly thereafter wrote to her brother Ezra in Ithaca, New York, to comment on the quality of her telegraph instruments and request further instruction on electricity:

> M. B. [her husband] told me to write to you to send me a Relay as good a one as could be got up. I cannot depend upon mine. He says it is a miserable thing. I am of course no judge. I think that I shall like tellegraphing if I have good instruments. I wish you would come and see us. I could now appreciate some instructions in regard to electricity.
>
> I hope to make a good Opperator soon[;] have no idea of trying to do the business unless I can learn to do it right. I have seen enough to know that a poor operator is a great source of annoyance.[6]

Phoebe Wood's comments on the quality of her telegraphic relay and the qualifications of her fellow operators are typical of an age when telegraphic instruments were largely experimental and handmade and all operators were more or less enthusiastic amateurs. Despite the embryonic state of the technology, however, Wood soon came to regard telegraphy as preferable to more traditional occupations for women in both pay and opportunity for self-improvement. Writing to her brother in November 1849 regarding another woman's occupational options, she expressed the view that telegraphy would enable women to improve not only their income but also their minds:

> Jane intends going to learn the tailor's work but I think she would do better to learn to telegraph. I hear they employ ladies in ofs. east. I cannot bear the thought of having her go from house to house to sew as Mary and destroy her health.
>
> It would not take nine hours in a day to earn in a telegraph office what she would have to work 10 hours with her needle, and in the former employment she would have time to improve her mind and keep her wardrobe in order.[7]

In the 1850s, women who understood the promise of the new invention and had acquired the necessary technical skills began to enter the field in increasing numbers. Helen Plummer became the telegraph operator in Greenville, Pennsylvania, around 1850; her brother P. S. Plummer delivered messages for her and repaired lines. Emma Hunter of West Chester, Pennsylvania, became a telegrapher for the Atlantic and Ohio Telegraph Company in 1851; Western Union would later designate her as the "first female operator," though it is clear that other women had worked as telegraphers before she did. Ellen Laughton became the operator in Dover, New Hampshire, in 1852 at the age of fourteen; she showed such proficiency at the work that she was promoted to manager of the Portsmouth, New Hampshire, office four years later.[8]

Surprisingly, the potential synergy between the railroad and the telegraph, the first two mass-scale technologies of the nineteenth century, was not remarked on for several years. Although early telegraph lines often

Figure 1. Emma Hunter, telegrapher, West Chester, Pennsylvania, 1851. From Reid, *The Telegraph in America*, 170.

Figure 2. Elizabeth Cogley, railroad telegrapher, Lewistown, Pennsylvania, 1855. From *Telegraph Age*, September 16, 1897, 382.

followed the rail right-of-way, no attempt was made to use the telegraph as a signaling system for the railroad until 1851, when Charles Minot, superintendent of the Erie Railroad, first used the telegraph to monitor and control the movement of trains.[9]

After the introduction of the telegraph for railroad dispatching, women began to work as railroad operators as well. Elizabeth Cogley of Lewistown, Pennsylvania, began to work for the Pennsylvania Railroad in 1855. She had been an operator for the Atlantic and Ohio Telegraph Company in Lewistown, Pennsylvania, and when the telegraph office at Lewistown was consolidated with the office of the Pennsylvania Railroad in the winter of 1855–56, she became the first known female telegrapher to work for a railroad. She had been employed as a messenger before she became an operator, a career path that was common for males in the nineteenth century but unusual for a female.[10]

Although the lack of census data on female workers makes it difficult to determine the numbers or percentages of women who worked as telegraphers in the 1850s, the names and life stories of a few women operators can be found in anecdotal accounts in the telegraph journals and newspapers. P. S. Plummer, reminiscing about the telegraphic career of his sister Helen in the pages of *Telegraph Age* in 1910, recalled that the operator in Conneaut, Ohio, in 1853 had also been a woman. *Telegraph Age* also mentioned that Sarah Carver became an operator in Fishkill Landing, New York, around 1857; the *Boston Herald* told how Nellie Reckards had begun work as a telegrapher in her uncle's store in Lynn, Massachusetts, in 1859, and served as operator there during the shoe workers' strike of 1860.[11]

Although the earliest women telegraphers en-

tered the profession for a variety of reasons, they shared a common understanding of the significance of the new invention and access to the skills required to operate it. Sarah Bagley's background in writing and journalism
made her aware of the possibilities of the telegraph, and her work with the
Voice of Industry put her in contact with the promoters of the telegraph in
Lowell, Massachusetts. Phoebe Wood entered the telegraph office in Albion, Michigan, both because of her brother Ezra Cornell's connection with
the telegraph business and the shortage of trained operators in that frontier
region. Emma Hunter also owed her position in part to kinship ties; the
telegraph line to West Chester, Pennsylvania, was built by a relative of hers,
Uriah Hunter Painter, who taught her the necessary operating skills. Elizabeth Cogley learned telegraphy from the previous Lewistown operator,
Charles C. Spottswood, who boarded with the Cogley family. Thus the
earliest women operators went largely unnoticed at a time when it was a
novelty to see persons of either sex operate the new and mysterious instruments.

The presence of women in the telegraph office was noted by Virginia
Penny, who included women telegraphers in her encyclopedic book *How
Women Can Make Money*. Although first published in 1870, much of the material was collected earlier, in the 1860s. Penny noted that there were approximately fifty women working in the northeastern United States for the
New York and Boston Magnetic Telegraph Company in the early 1860s,
some of whom were advancing beyond entry-level positions. As an example
of the possibilities of telegraphy as an employment for women, Penny reported: "In New Lisbon, Ohio, a young woman was employed a few years
ago, as principal operator in a telegraph office, with the same salary received
by the man who preceded her in that office." Penny noted that female operators entering the profession were beginning to encounter the "antagonism
naturally felt by male operators, who see in it a loss of employment to themselves." She also suggested that qualified women would have no trouble
outperforming their male counterparts: "Any female proficient in orthogra-
phy, with an inclination to useful employment, would make a good
telegraphist, and might readily command . . . a salary of from $300 to $500,
and be profitable to her employers beyond the ordinary male telegraphists
employed under the present arrangement of office."[12]

It was not until the end of the Civil War, when men began to return

from military service, that questions regarding the "proper place" and gender roles of male and female operators began to be discussed. While men
feared the loss of their livelihoods because women might be employed at a
lower rate, this prospect never materialized. Likewise, women were not
"driven from the trade," as some men recommended, or completely mar-
ginalized. Women continued to work in the telegraph industry after the
Civil War in part because of the support of the industry itself, and Western
Union in particular, but also because of the active efforts of women operators to defend and justify their role. Although women were predominantly
employed in lower-paying positions and in rural offices, women who persisted and made a career of the profession could work up to managerial
or senior technical positions that, except for wage discrimination, were
identical to those of their male counterparts. Telegraphy as an occupation
became gendered, in the sense that we understand today, only after the introduction of the teletype and the creation of a separate role for women
teletype operators.

WOMEN TELEGRAPHERS IN CANADA AND EUROPE

The development of the telegraph industry in Canada generally paralleled
the growth of the telegraph network in the United States. The two systems
were closely interconnected, in part because the submarine cables that first
connected North America to Europe in the 1850s and 1860s came ashore
in Canada. The telegraph industry in Canada was divided between private
telegraph and railroad companies, as in the United States, and the
government-operated Government Telegraph Service (GTS), which func-
tioned primarily in rural and maritime areas. Women operators were common after the mid-nineteenth century; they worked in both railroad and
commercial offices.[13]

In most European countries (and in most of the rest of the world), the
telegraphic system came under the control of the government posts and
telegraph administration rather than private companies, as in the United
States. Women first began to work as telegraphers in England for small private companies in the 1850s; after the telegraph system came under the
control of the British Post Office in 1870, increasing numbers of women
were employed as telegraphers. The telegraph administrations of the Scan-

dinavian countries and Switzerland also began to employ women in the 1850s. France, Germany, and Russia first admitted women to the telegraphic service in the 1860s. By 1900, women worked in all European administrations except for Bosnia-Herzegovina, Greece, Luxembourg, and Montenegro, where they were excluded from telegraphic service for religious or cultural reasons.[14]

WOMEN AND THE TELEGRAPH IN THE NON-EUROPEAN WORLD

The development of the telegraphic system in the non-European world in the late nineteenth century was closely tied to the rise of colonialism and the desire of the European powers to control their far-flung empires. The telegraphic system played a central role in the administration of the various territories, and undersea telegraph cables enabled direct connection between the European centers of power and the colonial governments.

Asia was connected to Europe telegraphically by both land-based lines and undersea cables in the 1860s and 1870s. Great Britain's British Indian Telegraph Company developed a system of undersea cables that ran through the Mediterranean to Suez and then through the Red Sea and the Persian Gulf to come ashore in what is now Pakistan in British India. To avoid dependence on the British-owned cable system for communications, the Siemens brothers in Germany built a land-based telegraph line that ran through Poland and Russia to central Asia. As the telegraph lines spread across the subcontinent, women telegraphers were employed in Japan, Ceylon, the Dutch East Indies, and French Indochina by 1900. In 1907, women were permitted to enter the telegraphic service in India and Burma (Mayanmar). Their employment was prohibited by law, however, in most Islamic countries, including Turkey and Tunis, and in Siam (Thailand).[15]

In 1871, the submarine cable from Calcutta in British India was extended as far as Australia; at the same time, an extensive network of overland telegraph lines was built in Australia to connect the isolated settlements. Women telegraphers were employed in the Australian provinces of South Australia, New South Wales, Queensland, and Victoria, as well as in Tasmania and New Zealand, by 1900.[16]

Africa was connected to Europe telegraphically by both the British undersea cables, which came ashore in Alexandria, Egypt, and by the French submarine cable system, which ran from Dakar in French West Africa (now Senegal) to Brest in Brittany. Extensive landline networks connected the cities of French West Africa and the British Cape Colonies (now South Africa). Women operators were employed in the British Cape Colonies, in the Portuguese colonies, and in French West Africa in 1900. Their employment was forbidden in Egypt.[17]

Mexico, South and Central America

Development of the Mexican railroad and telegraph systems had high priority under the dictatorship of Porfirio Diaz (1876–1911) and was often financed with foreign capital. Some women from the United States worked as telegraphers for the railroads in Mexico during the Diaz administration. One was Abbie Struble Vaughan, who went to work as an operator for the Mexican National Railroad after the death of her husband, J. L. Vaughan, in 1891. She remained in Mexico for over twenty years; when Diaz was deposed in 1911, she returned to Long Beach, California, where she taught telegraphy during World War I. Another was "Ma Kiley" (Mattie Collins Brite), who began her long career as a railroad telegrapher working in Sabinas, C. P. Diaz, Torreon, and Durango, Mexico, from 1902 to 1905.[18]

Chile developed one of the earliest telegraphic systems in South America after an American-born entrepreneur, William Wainwright, proposed building a national system in 1852. The Chilean telegraphic network was financed and administered exclusively by Chileans. The Chilean telegraph was also one of the first in South America to employ women as operators. Around 1870, a school to teach telegraphy to women was established at the National Institute, and soon thereafter women worked as operators in all parts of the country. Women also worked as telegraph operators in Argentina around 1900.[19]

Figure 3. "Ma Kiley" (Mattie Collins Brite), railroad telegrapher, Mexico, 1903. From *Railroad Magazine*, May 1950, 69.

WOMEN IN THE TELEGRAPH INDUSTRY IN THE TWENTIETH CENTURY

Use of the telegraph for sending commercial messages and routing trains reached its peak around 1900. The number of telegraphers began to decline in the early part of the twentieth century as the telegraph was replaced by the telephone for sending personal messages and by Centralized Train Control for routing trains. Although Morse instruments remained in service in remote parts of the world until the 1970s, they were largely replaced by Teletype devices for transmitting commercial messages.

The invention of the typewriter in 1867 led to the development of the "teletypewriter," or "Teletype," a telegraphic device with a typewriter keyboard that would transmit the character indicated by a keystroke. Sending and receipt of messages became largely automated; operators no longer had to transmit Morse code or switch circuits manually. Messages were typed on a standard typewriter keyboard at one end and printed out on a mechanical printer at the other. Various experimental Teletype systems were developed, beginning around 1900; telegraph companies were using them extensively by 1915.

Following the development of the Teletype, the functions of the telegraph operator began to resemble those of a typist. Messages were entered on a machine with a typewriter keyboard; at the receiving end, automatic printers printed out the text. A skilled operator who could decipher the Morse code dots and dashes was no longer needed. The introduction of the Teletype led to gendering of the occupation of telegrapher; almost all Teletype operators were women. Thus as the total number of telegraphers declined in the mid-twentieth century, the percentage of women employed in the industry increased.[20]

Daily Life in the Telegraph Office

LIFE IN THE DEPOT OFFICE

AN essay on the American telegraph industry that appeared in *Harper's Magazine* in 1873 noted that in much of rural America, the telegraph window at the railroad depot was the primary diversion for bored small-town residents: "A little group of loungers are gathered about the window-sill looking in." The object of their attention was the local telegraph operator; if the casual observer "should enter the room, he might see the operator at the key, holding a dispatch in her hand, and with the key making the strokes which are necessary for its transmission." Thus by 1873, women were so commonly employed as depot operators as to evoke little or no comment.[1]

Office environments were as varied as the operators who inhabited them. Nattie Rogers, the heroine of Ella Cheever Thayer's 1879 novel *Wired Love*, begins by enumerating the furnishings of her office: "a long, dark room into which the sun never shines, a crazy and a wooden chair, a high stool, desk, instruments,—that is all—oh! and me!" In contrast, Mildred Sunnidale, chief character of Josie Schofield's 1875 romance "Wooing by Wire," finds her office "pleasant and cozy"; she has put down a carpet, covered up the inkstains with green baize, and placed a geranium in the window. After only a week of her presence, the office already wears "a tidier and more cheerful aspect than it used to under the old masculine regime."[2]

As the telegraph was increasingly used in railroad management and telegraph stations could be found in nearly every local railroad depot, the telegraph companies saw an opportunity to increase their business by developing a cooperative strategy with the railroads. The telegraph companies installed and maintained the lines in the right-of-way provided by the railroad and supplied the instruments in the depot. The railroad paid the wages of a telegrapher, who would handle not only train messages but also personal and commercial telegraph messages. Thus the railroad got a signaling system maintained by the telegraph company, while the telegraph company got stations in many small towns and rail crossings where one might not otherwise have been economically feasible.

Although operators frequently altered office furnishings to suit their personal tastes and express their individuality, the work was the same for men and women. There were telegrams to be sent and received and train orders to copy and "hand up" to passing trains.

There were many different classes of telegraphic messages to be handled. A "day letter" was the standard-priority daytime telegram; a "full

Figure 4. Terminal station, 1873. From *Harper's New Monthly Magazine*, August 1873, 332.

rate" message was a higher-priority "rush" message. (In telegraphic terminology, a "letter" had a fixed length, say fifty words, and a flat-rate charge,
while a "message" was of unspecified length and was charged for by the
word.) A full rate message had priority over everything except government
messages, which had the highest priority of all. "Night letters" and "night
messages" were less expensive because they were sent at night, when the
line was less busy. There was also less chance that the operator at the receiving end would be awake and at the key so a night letter might not be
delivered until the next day, and a night message would definitely not be delivered until the next day.[3]

Figure 5. Office of Carrie Pearl Seid, railroad operator, Sunbury, Pennsylvania, 1907.
From *Railroad Telegrapher*, August 1907, 1256. Reproduced from the Collections of the
Library of Congress.

Anne Barnes Layton, reminiscing in the 1950s about her work as a telegrapher at the Woods Cross, Utah, railroad depot in the early 1880s, remarked that customers did not always carefully count the words in the telegrams they handed her, thus requiring her to edit the messages as she sent them: "In those days you could send ten words in a telegram for a flat charge, instead of the fifteen words now permitted. When the train stopped at Woods Cross, the passengers would run to the telegraph office, write their telegrams, pay the flat rate, and rush back to the cars. I didn't have time to count the words until the trains had gone. Then I had to do a lot of condensing, but I never omitted the essentials."[4]

Railroad telegraph operators also received train orders from a dispatcher and passed them to the train crew. A train order might require a train to speed up or slow down so it would arrive at the next station at the proper time, halt at a siding to allow a "superior" train coming from the other direction to pass, or inform the engineer of an unscheduled stop to pick up freight or passengers.

The dispatcher controlled the movement of trains from a central location and telegraphed train orders to each order station on the line. Most dispatchers were men, but a few were women. Medora Olive Newell learned telegraphy in the Chicago Great Western depot at Durango, Iowa, at the age of fourteen; she served as a dispatcher in Dubuque and Des Moines, Iowa, during the 1890s. Rebecca S. Bracken was a dispatcher for the Michigan Central Railroad in Niles, Michigan, for more than forty years before her retirement in 1905. While male dispatchers, at least as depicted in railroad fiction, were often gruff and dyspeptic characters who kept operators in constant terror at the prospect of an incorrectly copied train order, Rebecca Bracken was remembered for her pleasant personality. According to her obituary in *Telegraph Age*, "She was a woman of lovely character, and was known as 'the angel' by the railroad men."[5]

Telegraphers in turn had little respect for a dispatcher who issued vague or misleading train orders. Train orders were intended to be literal, step-by-step instructions for the movement of trains; an error in wording might result in delays, collisions, and loss of life. Ma Kiley brought a serious error to the attention of a dispatcher while she was working for the Southern Pacific in Nevada around 1940: "I kept a copy of one of his orders for quite a while, an order which read 'Extra so-and-so west remain on siding at Fernley and meet so-and-so at Thisbe.' When I called him on it he wanted to know what was wrong. I asked him how the h—— could a train remain on siding and meet another train miles and miles from the place designated!"[6]

A "number 31" train order required the train to stop and sign for the order; a "number 19" order had to be "hooped" or "handed" up to passing trains, using a five-foot-long order staff. Handing orders could be dangerous if not performed properly, as Sue R. Morehead recalled in *Railroad Magazine* in 1944:

> Then I prepared to go out and hand up.
>
> I had literally no idea how this should be done. Common sense told me that I should have a light of some kind. I found an old hay-burner lantern, of the type car inspectors use to throw a spot.
>
> The train was nowhere near the station when I went out and placed myself entirely too close to the track, lantern at my feet and hoop in the air. By the time the extra got close, my arm ached and my whole body trembled. The big 5000 coming toward me with its headlight shining in my eyes loomed larger and larger; I had a moment of paralyzing fear, when every instinct told me to light out from there and keep going. . . .
>
> I held the train order hoop tightly, and only the fact that I stood too close and the brakeman missed the hoop kept me from being pulled into the train. The head man came back after the train stopped and took the order. With amazement, he advised me not to stand so close and to hold the light up so that the hoop could be seen.[7]

MORSE KEYS AND "BUGS"

Operators used different keys to send Morse code. During the nineteenth century, telegraphers used the traditional Morse telegraph key, invented by Samuel Morse and Alfred Vail in the 1840s. The standard Morse key had a spring-loaded movable lever, suspended between two pivot points, with a hard rubber knob which the operator grasped between her thumb and first two fingers to operate. A small up-and-down motion was

Figure 6. Handing orders, *Railroad Stories* cover—1930s. From "Women and the American Railroad" Collection by Shirley Burman; reprinted from *Railroad Stories* (now *Railfan and Railroad Magazine*).

Figure 7. Standard Morse key. From Pope, *Modern Practice of the Electric Telegraph*.

required to make and break the circuit. There were several different styles of Morse telegraph keys; a "camelback key" had a characteristic "hump" in the lever, while most keys had a straight lever.[8]

Around 1900, the Vibroplex or "bug" automatic sending key became the instrument of choice for low-strain, high-speed sending. The bug's lever moved sideways, rather than up and down, as did the original telegraph key. Weighted contacts attached to either side of the lever caused the key automatically to produce a series of dots if the lever was moved in one direction and a series of dashes if moved in the other. This reduced the strain on an operator's arm when performing fast sending and prevented "glass arm," a temporary paralysis caused by long periods of high-speed transmission. The bug came into common use around 1900; the Vibroplex version was patented in 1904. Many operators owned their own bug, which

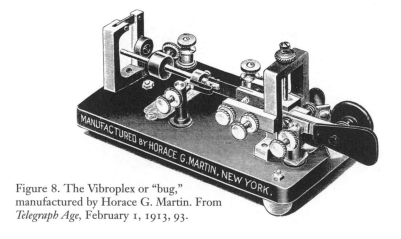

Figure 8. The Vibroplex or "bug," manufactured by Horace G. Martin. From *Telegraph Age*, February 1, 1913, 93.

they took with them when they moved from one job to another; it became a visible symbol of first-class status. Ma Kiley, a career telegrapher, became so attached to her bug that she titled her autobiography, which appeared in *Railroad Magazine* in 1950, "The Bug and I." As she stated, "That bug and I really went places. It enabled me to become what was termed a first-class Morse telegrapher, on the railroads and for the Western Union, something that will soon be alive in memory only because the teletype and other machines have put us on the shelf."[9]

LIFE IN THE COMMERCIAL OFFICE

In larger towns, the volume of telegraphic traffic quickly outgrew the capabilities of the depot operator, and a commercial telegraph office was opened, typically in the business district of the city, to provide fast access to banks, businesses, and newspaper offices. In this environment, anywhere from a few to a few hundred operators might work in a hierarchically structured work environment in which managers supervised operators and operators supervised clerks and messengers. In the larger offices, the work

Figure 9. Postal Telegraph office, Cincinnati, Ohio, 1907. From *Commercial Telegraphers' Journal*, October 1907, 1036.

of the telegraph operators was further stratified. A chief operator supervised technical operations, first-class operators handled press reports and market quotations, and second-class operators handled local traffic and personal messages.[10]

Duties of the Manager

The manager of a telegraph office oversaw the general operations of the office, which included hiring and firing of personnel, payroll, recordkeeping, assignments and promotions, and liaison with corporate headquarters. Although most managers were male, it was not uncommon for women to manage a commercial office.

Typically the manager was a telegrapher who had come up through the ranks, as operator and then chief operator. Hettie Ogle, manager of the Johnstown, Pennsylvania, Western Union office in the 1880s, started as the sole operator in 1869 and rose through the ranks as the office grew and new personnel were added. She maintained a low profile about her gender, signing official correspondence as "H. M. Ogle"; many at Western Union thought the Johnstown manager was a man.[11]

Some women found that a willingness to relocate led to better opportu-

Figure 10. Hettie Ogle, manager, Johnstown, Pennsylvania, Western Union Office, 1889. From McLaurin, *The Story of Johnstown*, 178.

nities. Fannie Wheeler started out as a telegrapher in Blairstown, Iowa, and eventually moved to Chicago, where she became manager of the ladies' department in 1869.

The ladies' department in Chicago was actually the city department, the division of the telegraph office that was responsible for personal messages and local traffic. The city department generally included not only the operators at the main office but also those at branch offices in hotels and stores throughout the city. Many city departments became predominantly female in the 1870s; male operators who ventured into the city department in those years often became the object of teasing and "hazing" from the women. In the city department in San Francisco in 1875, for example, according to the *Telegrapher*, the five women operators "all contribute toward keeping the only gentleman, Mr. John A. Smith, at the Market Street office, as embarrassed as a young lady without pins."[12]

Duties of the Chief Operator

The chief operator was the senior technical person in the office and was responsible for the proper operation of the lines and equipment and assigning tasks to the subordinate operators. Chief operators had to be familiar with all technical aspects of telegraphy and had to solve any problems with the wires and equipment. Women chief operators included Hettie Ogle's daughter Minnie, chief operator at the Johnstown office of Western Union in the 1880s, and Elizabeth Cody, chief operator of the Springfield, Massachusetts, Western Union office in 1905. A *Telegraph Age* feature on Elizabeth Cody noted that

> the chief operator is called upon to test all trouble on wires within a radius of twenty-five miles of the city on all of the various routes and if trouble is found to see that it is immediately remedied. She also has to look out for the traffic of all the telegrams and see that they are sent out promptly and she has charge of the alarm and night watch service which is an important feature. By this system safes such as that in the office of the city treasurer are safeguarded by wires and the moment the safe is tampered with in the least way announcement is made by a big gong ringing at the telegraph

office and the chief operator hurries a man post-haste to see what the trouble is.

The *Telegraph Age* article described the Springfield Western Union office as a fairly typical medium-sized commercial office at the turn of the century. It was on the main line between New York and Boston, and the 120 telegraph lines that passed through the office handled a large share of the brokerage traffic. Mrs. Cody and her staff of eleven operators, five clerks, and fifteen messengers handled approximately one thousand messages a day.

Although *Telegraph Age* described Mrs. Cody's position as "an unusual one for a woman," her background and training were typical for senior-level telegraphers. Elizabeth Cody entered telegraphy after graduating from high school in 1892; her on-the-job training and twelve years' experience qualified her for the top technical position.[13]

Duties of Operators

Operators were responsible for transmitting and receiving messages in Morse code. To receive a telegram, an operator would copy the text onto a blank message form as she decoded the clicking of the sounder. After copying the message, the operator would proofread the message and verify the word count against a "check" value contained in the message. The time of

Figure 11. Elizabeth A. Cody, chief operator, Springfield, Massachusetts, Western Union Office, 1905. From *Telegraph Age*, May 1, 1905, 180. Reproduced from the Collections of the Library of Congress.

receipt would then be marked on the message, and receipt of the message would be noted on a log of received messages. The completed message would then be handed to a delivery clerk or directly to a messenger.

Good handwriting was an important skill for early telegraphers because messages had to be copied onto message blanks as they were received. Telegrapher's script, an elaborate form of cursive handwriting that connected the letters in a continuous line, was the preferred form of transcribing messages. Some offices allowed operators to make a first copy in pencil and then trace over it in ink after correcting any mistakes; this practice was forbidden in fast-paced, high-traffic offices like the central office of Western Union in New York City.

Around 1890, the typewriter was introduced into the telegraph office, and operators were allowed to type messages directly onto message forms as they received them. The typewriter quickly became the predominant means for copying messages; it proved to be a boon to those whose careers had been hindered by deficient handwriting skills. The typewriter was in turn replaced by the Teletype, beginning around 1915.

Novice operators often missed a few letters of an incoming message; to request retransmission, they would operate the switch lever and "break"

Figure 12. Typewriter ad, 1912. From *Telegraph and Telephone Age*, July 16, 1912, 480.

into transmission, much to the annoyance of anyone else trying to receive the message. Thus new operators were under pressure to come up to speed quickly and avoid time-wasting and embarrassing requests for retransmission. To transmit an outgoing message, the operator had to open the switch lever, which shorted out the key during reception. The key could then be used to send; only one operator could send on a standard line at any time.

The operator would send the text from the message form filled out by the customer. After completing the transmission, the operator would write her "sign" (a name or initials), the sign of the distant receiving office, the time of transmission, and the "check," which indicated word count, on the message form. A notation would then be made on an outgoing message log and the original filed in a transmitted message file.

It was important to have a smooth, graceful sending style; beginners were often accused of sending in an awkward, choppy style. "Clipping" was an affected sending mannerism in which the proper duration was not given to each dot or dash; male telegraphers frequently accused women of clipping.[14]

Many telegraphers experienced "stage fright" the first time they were called upon to send a message. Barbara Gowans, a Tooele, Utah, operator in the 1870s, had an experience that was typical for many first-time operators: "I will never forget the first message I sent. . . . I was very nervous, I tried but all that was received was the address and the signature."[15]

An operator's "sign" was an important part of her identity; it signified membership in the telegraphic fraternity. Emma Hunter of West Chester, Pennsylvania, was better known to other telegraphers as "Emma of S." Women operators sometimes made up fanciful names to use as their personal signs so as to create an atmosphere of mystery or just to be playful. In Utah in 1867, Mary Ellen Love, Elizabeth Claridge, and Elizabeth Parks, the first three women to become operators for the Deseret Telegraph, signed themselves as "Estelle," "Lizzette," and "Belle," respectively. Their signs stayed with them as nicknames for years; even their families began using them. Women often used their signs when writing to the telegraphic journals, particularly regarding women's issues; use of a sign allowed a telegrapher to identify herself to other operators while still preserving personal anonymity.[16]

To qualify as a first-class operator, one had to be able to send and receive error-free code at thirty to forty words per minute. Typically, five years of experience were required to become a first-class operator. First-class operators were assigned the more critical business traffic and press reports. Although this work paid better, it was boring and tedious, consisting of little more than nonstop sending and receiving at high speed.

Mary Macaulay worked as a press operator in Auburn, Rochester, and Syracuse, New York, before becoming vice-president of the Commercial Telegraphers' Union of America (CTUA) in 1919. Minnie Swan's career took her from the Cincinnati office of Western Union to the New York offices of the Baltimore and Ohio Railroad, and finally to a coveted position at a brokerage office, one of the highest-paying jobs available, in the early 1880s.[17]

Figure 13. Mary Macaulay, vice-president, Commercial Telegraphers' Union of America (CTUA), 1919. From *Commercial Telegraphers' Journal*, November 1919, 529. Reproduced from the Collections of the Library of Congress.

The first rung on the operator hierarchy was the entry-level position, normally referred to as a second-class operator (although some telegraph companies had as many as four grades). Typically a second-class operator could receive and transmit at ten to twenty words per minute with relatively few errors. He or she would be assigned the less demanding and slower work such as personal messages and local traffic. Most women operators in the nineteenth century were second-class operators.

Clerks

Clerks, also called "check-girls" or "check-boys," were responsible for maintaining the flow of messages between the customers and the operators. In large offices they would accept messages from customers, divide them up among the operators, and deliver them to the operators for transmission. They would also collect received messages from the operators and give them to the messenger boys for delivery. Being a clerk was a stepping-stone into the operator position; many operators started out as check-boys or check-girls.

Messengers

Messengers delivered telegrams to customers. Messengers were mostly young boys, but a few girls and women delivered messages, mostly in small towns and rural areas. Elizabeth Cogley started out delivering messages in Lewistown, Pennsylvania, in 1852, as did Virginia Brom in Pella, Iowa, ninety years later. Messengers generally received a small salary but subsisted mainly on tips.

THE WORKING ENVIRONMENT

In some of the large telegraph offices in metropolitan areas of the United States, operators were segregated by sex during the mid-nineteenth century. This was done in part to reassure anxious fathers and husbands that the working environment of their daughters and wives would not be corrupted by the presence of male operators, who had acquired a reputation for foul

Figure 14. Clerks and messengers, Raleigh, North Carolina, 1912. Courtesy Western Union Telegraph Company Collection, 1848–1963, Archives Center, National Museum of American History, Smithsonian Institution, SI neg. #94-1819.

language, tobacco and alcohol use, and a generally dissolute lifestyle. The ladies' department was off bounds to men, as wryly noted in a ditty that appeared in the *Operator* in 1876:

> *Nor must you ever wildly stare*
> *To gaze on the ladies over there,*
> *For if you do, you may safely swear*
> *You'll get reported and "bounced" then and there.*[18]

In smaller offices, partitions and even cabinets were occasionally built to ensure privacy for the female operators. The idea of the "ladies' compartment" was used to provide a humorous touch to a telegraphic romance of 1876, "The Thorsdale Telegraphs," which appeared in the *Atlantic Monthly*. Jahn Thor, the stationmaster, sensitive to notions of propriety, has partitioned the telegraph office and created a private compartment with a locking door for his new assistant, Mary Brown: "I peeped into a kind of closet some six by eight, and ventured in, when he shut the door softly after me. It was the merest bin of a place, roughly put together, though there had been apparently quite a determined effort to make it look comfortable. There was a stone pitcher, a tin basin, a looking-glass, and a pretty bunch of flowers in an ale bottle."[19]

Intentionally or not, this segregation by sex became segregation by skill level as well. Lewis H. Smith, editor of the *Telegrapher*, noted in 1865 that "teaching a young lady the rudiments of the business and then cooping her up in a room by herself or with others of her sex, away from all chance of gaining knowledge, or emulating those who are in the front rank," would cause her to lose opportunities for advancement and create a permanent "underclass" of female operators.[20]

This segregation did not occur in England, where male and female operators generally worked together. The postmaster general commented favorably on this arrangement in 1871, stating that "there has been no reason to regret the experiment." An additional benefit from this arrangement, in the view of the postmaster general, was that "it raises the tone of the male staff by confining them during many hours of the day to a decency of conversation and demeanour which is not always to be found where men alone are employed." Further, he noted that "it is a matter of experience that the male clerks are more willing to help the female clerks with their work than to help each other; and that on many occasions pressure of business is met and difficulties overcome through this willingness and cordial co-operation."[21]

In the Chicago office of Western Union, segregation by sex was abandoned in 1871, shortly before the office on LaSalle Street was destroyed in the Chicago fire of 1871. The *Telegrapher* reported in its November 18, 1871, issue that "the separate female telegraph department of the office was abolished, the communication between the rooms thrown open. . . . This was a most sensible change, and one which might be profitably and advantageously adopted in other large offices. The absurdity of cooping up operators, because they chance to be women, in a room by themselves, is one that should long ago have been abandoned."

The editors of the *Telegrapher* added, "If women are to be telegraph operators at all, they should be placed upon the same basis as men who fill similar positions, and should, for equal ability and equal positions, receive as much pay as their male associates." An extra benefit for the women operators, in the view of the *Telegrapher*'s editors, was that it would "relieve them from the tyranny which is apt to be exercised by some favorite of the Superintendent, who is made 'Manager of the Ladies' Department,' and who, being isolated from the rest of the office, is enabled to render their existence miserable, if not absolutely intolerable."[22]

The last remark is a thinly veiled allusion to Lizzie Snow, manager of the city department at Western Union in New York; stories of her dictatorial conduct had already reached the ears of the editors of the *Telegrapher*. She would be dismissed by Western Union for "refusal to submit to and obey certain rules and regulations of the office" in 1875.[23] Western Union continued to maintain a female-only city department in its main New York office, under the more lenient direction of Snow's successor, Frances Dailey, until the 1890s, although it gradually removed the physical barriers that separated women operators from their male counterparts.

When the Western Union operating room was moved from 145 Broadway up the street to 195 Broadway in the spring of 1875, the 150 male operators and 60 female operators were placed in the same room on the seventh floor. The new operating room was attractively decorated; according to the

Telegrapher, "the side walls of this room are tinted a faint lilac color, the ceiling being bordered with a tasteful and elaborate tracery, the prevailing colors being buff, blue, and gold." The *Telegrapher* noted, however, that the operating room was "sadly marred by a straight, ugly screen stretched across the room about two thirds of the way down, which cuts off the perspective of the row of tables, and entirely breaks up the unity of artistic effect." This screen was intended to separate the women's working environment from that of the men: "There seems to be no necessity whatever for such a barrier between the ladies department and the remainder of the room, and the management would display good sense, as well as vastly improve the appearance of the apartment, if they would abolish it altogether, as is done in most other large offices."[24]

Sometime before 1881 the barrier was apparently taken down; in that year a telegrapher, Ambrose Elliott Gonzales, wrote home to South Carolina:

> The office is in the 7th story of a large & beautiful building & we go up in an Elevator, a kind of large box that is hauled up to the top of the house by steam. The operating room is 250 feet long & about 70 wide so you may know what a big Office it is. There are about 300 Men employed & about 75 Girls & women in one room as operators, think what a noise seven hundred sets of instruments must make. Some of the girls are quite pretty but I don't waste much time on them.

Ambrose Gonzales makes no mention of a divider; he evidently could see the women from where he sat, although he professed to ignore their presence. In any event, the illustration of the main operating room at 195 Broadway which appeared in the July 1889 issue of *Scribner's Magazine* shows men and women working side by side.[25]

Integrating the work environment required providing separate bathroom and "retiring room" facilities for female employees. In the railroad depots, women operators simply made use of the facilities provided for female travelers, usually primitive outhouses; in the larger offices, bathroom facilities had to be maintained, and women employees had to be given access to them during working hours. Male managers were occasionally reminded that female employees had to deal with menstrual periods, as in an editorial in the April 15, 1882, issue of the *Operator:*

> It is true, however, that women in the profession have suffered much from the thoughtless negligence or wanton lack of consider- ation of employers in sanitary matters, of which delicacy has for- bidden their making complaint. It is the duty of every telegraph superintendent or manager who employs women to see that they are provided with suitable retiring rooms, with sufficient privacy, and are afforded opportunities to visit them without notice.[26]

Although not considered a topic of polite conversation during the nine- teenth century, the condition and cleanliness of bathrooms became an issue in the strike of 1907. Women operators in Chicago placed the statement that their "withdrawing rooms and conveniences are a disgrace to human- ity" near the top of a list of grievances included in an open letter to Helen Gould, a Western Union stockholder.[27]

Figure 15. Western Union main operating room, 195 Broadway. From *Scribner's Magazine*, July 1889, 8.

WORKING HOURS

Working hours depended on the type of office. The typical working day at a large Western Union office in the 1880s was ten hours; since the office was in continuous operation, operators took turns doing evening and Sunday "tricks," or shifts. Evening shifts were shorter, typically seven and a half hours. No premium was offered for evening or weekend work; this became an issue in some labor disputes. By the early twentieth century, telegraphers had demanded, and gotten, a fifty-four-hour week for men, and a forty-eight-hour week for women.

Women were generally not expected to work at night because it was not considered proper for respectable women to be out alone at night. Also, before the invention of the automobile, it would have been difficult for many women to arrange transportation. Requesting women to work nights was reported as an example of "abuse" before the U.S. Commission on Industrial Relations in 1915.[28]

Women telegraphers did work nights in many circumstances. Railroad operators rarely had the luxury of regular working hours; they had to be present whenever trains passed the station. Late-arriving trains forced operators in single-operator offices to remain after dark. In some cases, women were given permission to work nights to make up for time lost as a result of illness or other absence; this warranted a mention in the *Telegrapher* in 1876:

> It has never been the custom for any of the Chicago lady telegraphers to work nights, but I understand that recently the W.U. ladies appealed to the manager of that office to allow them to make up the lost time deducted on account of sickness, etc. by working extra now and then, as the gentlemen do. His consent was finally obtained, and one of the ladies began working two or three nights a week. Saturday night, however, the night hawks were agreeably surprised to see another one of the ladies, one of the best operators in the company's employ here, on hand with the former mentioned lady, ready for extra duty."[29]

In England, postal regulations limited the working hours for women operators in 1900 to the period between 8 A.M and 8 P.M., although, in the words of Charles Garland, their "inability to perform night duty" was no more than "the result of an artificial social prejudice"; he noted that no one objected to women working as night nurses in hospitals. In Paris in the same year, women employees of the telegraph office worked from 7 A.M. to 9 P.M., and in Berlin they could be employed until 10 P.M. The only European administration that placed no restrictions on night work for women in 1900 was that of the Netherlands.[30]

Women operators were expected to take their turn at working Sunday shifts; this caused some distress to the deeply religious. Many companies allowed those who preferred not to work on the Sabbath to hire a substitute. The April 15, 1882, issue of the *Operator* contained a letter from "Mizpa" complaining that the Baltimore and Ohio Telegraph Company had suspended the practice of allowing women to hire substitutes on Sunday and now required them to work their assigned shifts.[31]

Working hours at single-operator offices were subject to the operator's availability, although most kept standard business hours. In a single-operator office, the operator might have days when almost no work was required and others that required fifteen hours straight at the key. Some of the most desired jobs were at brokerage houses, where telegraphers handled stock and commodities quotations. The working day was short, typically five and a half hours.

Around 1910–15, many states passed laws limiting working hours for women. Through the combined influence of reformers, labor unions, and the National Consumer's League, state governments set statutes that limited the working day to eight hours for women. While these laws were beneficial for many women working in factories, they sometimes made it difficult for women operators to get employment with railroads, where working hours were long and unpredictable.[32]

INTRODUCTION OF THE TELETYPE

The Teletype was a fully automated telegraphic sending terminal used for typing messages for transmission to a remote location; it first began to be extensively employed in large commercial offices around 1915. The name is simply a combination of "telegraph" and "typewriter"; the machine enabled an operator to transmit messages by simply typing the message on a type-

writer keyboard. In earlier versions of Teletypes (the multiplex), the Teletype would output a punched tape containing the text in Baudot code, the
ancestor of modern computer ASCII code. (Another early Teletype, the
Morkrum, was able to send messages directly without requiring an intermediate punched tape.) The punched tape would then be fed into a multiplex-
ing unit for transmission and was merged with the data of as many as eight
other messages and transmitted as a common data stream to a remote location. There, the stream would be demultiplexed and each message fed to an
automatic printer that printed it out either on a sheet of paper or on paper
tape. Later Teletype units transmitted data directly, eliminating the need for
the paper tape.[33]

Morse operators saw the Teletype as a threat because it did not require
skilled telegraphers to operate it. A veteran telegrapher who lost his job to a
Teletype machine in 1917 described the new device as a "efficient mechanical

Figure 16. Multiplex operating department, 60 Hudson St., 1930s. Courtesy Western
Union Telegraph Company Collection, 1848–1963, Archives Center, National Museum
of American History, Smithsonian Institution, SI neg. #89-12929.

Satan" that drowned out the sound of his Morse sounder and "delivered yard after yard of clean printed copy in a jerky, steam-roller voice—a voice that could, however, automatically and without fatigue, forever announce the typewritten words on an endless roll of smooth white paper."[34]

Since the speed at which messages could be sent was limited only by the typing speed of the operator, messages could be sent at a phenomenal rate of speed compared to using a Morse key. While the average rate of transmission using a Morse key was approximately 60 messages an hour in the New York Western Union office in 1917, one operator, Lillian Wagenhauser, was able to send a phenomenal 173 messages in one hour using a Morkrum Teletype unit.[35]

While Ma Kiley was correct in stating that the Teletype put many traditional Morse operators like herself "on the shelf," it also led to a large increase in the employment of women in the telegraph industry. Many of the Teletype operators were women employed at a lower rate than their male counterparts. Wide-scale use of the Teletype beginning around 1915 greatly increased the number of women employed in the telegraph industry

Figure 17. Printer department operators on break, Dallas, Texas, 1912. Courtesy Western Union Telegraph Company Collection, 1848–1963, Archives Center, National Museum of American History, Smithsonian Institution, SI neg. #94-1818.

because a female Teletype operator could send and receive at a rate far exceeding that of a first-class Morse operator, and at a lower rate of pay.

OCCUPATIONAL HAZARDS

Although telegraphy was less physically strenuous than most factory work in the nineteenth century and generally conducted in more healthful surroundings, it entailed some risks. Lightning was an occupational hazard for telegraphers; when a bolt of lightning struck the wire, the electrical charge would travel along the wire to the telegraph office, where it would take the shortest path to ground—sometimes through the operator. Irene Van Alstine, an operator in Akron, Iowa, around the turn of the century, was copying a weather report in the depot office when lightning struck the wire and knocked her out of her chair. She was uninjured but discovered burn marks on her body underneath the steel stays in her corset, which had evidently conducted the charge.[36]

Lizzie Clapp, an eighteen-year-old operator for the Boston and Providence Railroad at Readville, Massachusetts, was less fortunate; she was killed by lightning as she sat in the window of her station during a storm in 1876. The clergyman who officiated at her funeral alluded to the fact that her death had been caused by electricity, "an agency which she had been accustomed to guide and control." The death of Lizzie Clapp struck a special chord in telegraphers, even beyond the standard Victorian-era penchant for morbidity; the case was discussed almost obsessively in the journals for months. One outcome was a renewed interest in safety; it was pointed out that the Readville office, like many others, had not been properly grounded, and no one should have been near the telegraph equipment during a storm. The use of telegraphic lightning arrestors, beginning in the 1870s, reduced this hazard, but it was still best to ground the line when it was not in use and to stay away from the apparatus during a storm.[37]

Disease also took a toll among operators. Consumption, or tuberculosis, was often cited by the telegraphers as an occupational hazard because they spent long hours in poorly ventilated surroundings. Narciso Gonzales remarked in 1877 that "having to go out of a steaming hot office into the freezing air outside" at the Savannah, Georgia, office of the Atlantic and Gulf Railroad was "an excellent opportunity for my old hobby,

—consumption—to step in." Although Gonzales managed to remain in good health, tuberculosis claimed the life of Chicago telegrapher Annie Casey Tipling in August 1893 after a futile trip to the South in search of a healthier climate.[38]

Although poor ventilation and a sedentary lifestyle may have contributed to a high incidence of tuberculosis, a more common cause of illness or death was typhoid fever carried by contaminated water supplies. Twenty-six-year-old Josie C. Adams died of typhoid fever while working as an operator for Western Union in Detroit, Michigan, in 1876, and railroad telegrapher Ma Kiley nearly succumbed to the same disease at the age of twenty-two in 1902 while working in Sabinas, Mexico. As she recounted in her autobiography, "The Bug and I," "I got sick but was afraid to admit it or lay off for fear I'd never get another job. I held on until a train crew found me unconscious in the office one night. I never knew how I reached the hotel. After that I took the train for Del Rio and stayed there 90 days, for I had a bad siege of typhoid pneumonia and came very near death."[39]

A less serious but common affliction among telegraph operators was "glass arm" or "telegrapher's cramp," now known as "repetitive motion syndrome" or "carpal tunnel syndrome," caused by the nature of the hand operation used to send Morse code continually at high speed. Martha Rayne described the condition in 1893:

> There is the disease known as telegraph cramp, the diagnosis of which has not yet been thoroughly ascertained by the physicians. An operator stretches out her hand to press her finger upon the button of the instrument, and suddenly her arm refuses to obey her will, and lies numb on the desk beside her. If the tendons of her wrist had been cut through, her manual helplessness would not be greater. The strongest voluntary force is too feeble to make itself felt at the ends of the fingers. The operator simply can not do her work.[40]

The only known treatment for the affliction was to stay off the key until the symptoms disappeared. In the 1890s, advertisements for arm braces began to appear in telegraphic trade magazines. Introduction of the Vibroplex or "bug" around 1900 greatly reduced the strain involved in sending and thereby lowered the incidence of glass arm.

The work of the telegraphers required a complex set of business and technical skills, especially by nineteenth-century standards. In addition to Morse code, the operator had to adjust and maintain the telegraphic apparatus, do some typing, bookkeeping, and filing, and often perform railroad signaling as well. While a hierarchical organization existed in the telegraph office, with senior operators performing the more complex technical functions, women were not excluded from any of the positions held by their male counterparts, though they were frequently paid less and had fewer opportunities for advancement. Thus the presence of women in technical and managerial positions in the telegraph industry contributed to a perception of women operators as "exceptional."

Although the depot telegraph office continued to exist well into the twentieth century, the increase in telegraphic communication and the growing industrialization of the profession led to the development of a new work environment—the commercial telegraph office—which resembled a modern corporation in its organization and operation. Several different strategies were used to assimilate women into this new work environment, including physically separating women from the male workforce. By the 1880s, however, men and women Morse operators worked together in a common work environment. It was only with the introduction of the Teletype in the early twentieth century that truly gendered roles developed in the telegraph industry.

Society and the Telegraph Operator

T ELEGRAPHERS were among the first of the technological elites to appear in the mid-nineteenth century; to small-town people, they were modern Prometheans who knew how to make the electric wire talk. Men and women telegraphers alike were regarded with awe; they were the first members of what Marshall McLuhan would later call the "Global Village," and they connected their local communities to a larger world of commerce and politics. The world of the telegraphers encompassed all the places connected by the telegraph wire; they listened to elite gossip from all over the world. Although Nattie Rogers of *Wired Love* might find her office "dingy and curtailed," nevertheless it was a place from whence she could "wander away, through the medium of that slender telegraph wire, on a sort of electric wings, to distant cities and towns." Minnie Swan, reminiscing in 1937 about her telegraphic career fifty-five years earlier, recalled that "it meant something, in those . . . years, to be a telegraph operator. They were looked upon with wonder as possessing knowledge which separated them from the rest of the crowd."[1]

SOCIAL CLASS

The aura that surrounded telegraphers in the mid-nineteenth century was in part owing to their ambiguous social standing; telegraphers occupied a new

and almost indecipherable niche in the social hierarchy. They were "information workers" and "technicians" well before these familiar late twentieth-
century job classifications existed, making attempts to discuss them in
nineteenth-century terms particularly elusive. As Edwin Gabler notes in his
1988 study *The American Telegrapher,* "Nothing comparable to telegraphy
existed before the mid-nineteenth century."[2]

For most operators in the United States, telegraphy was a social means,
rather than an end in itself; the occupation was often presented in the trade
journals and in the reminiscences of telegraphers as a means of advancing or
improving one's station in life. The names of such famous men as Thomas
Edison and Andrew Carnegie were often invoked as examples of those who
had risen from humble beginnings to exalted positions thanks to time spent
in the telegraph office. Though less frequently and less visibly, women
telegraphers also shared in this Horatio Algeresque vision, as when Western
Union's *Journal of the Telegraph* referred in 1869 to Lydia H. (Lizzie) Snow,
manager of the City Department of the New York Western Union office, as
"a competent instructor . . . who had risen by steady steps to the very front
rank of American telegraphers."[3]

Like their male counterparts, women telegraphers were an important
part of the new upwardly mobile lower middle class. The term "middle
class," though, had vastly different connotations for its members in the mid-
nineteenth century than it does today. For them, the term implied newness
and a modern outlook; for women in particular, it suggested the possibility
for education, independent living, and meaningful work outside of the
home.

Alexander H. Bullock, former governor of Massachusetts, conveyed the
mid-nineteenth-century sense of this term, especially as it applied to
women, in a speech he delivered at Mount Holyoke Seminary in 1876, entitled "The Centennial Situation of Woman":

> In periods when there were only the gentlewoman and the low-
> born woman, the one was indeed maintained by the other, but
> the one also belonged to the other, or to the master of both; and
> self-dependence, whether ideal or actual, was as unknown as
> the electric telegraph. In the progress of time the uprising of a
> middle class, and the introduction of shop-keeping and textile

manufactures, stimulated the dead level of female life; and in the subsequent growth of this middle class, which in every nation has come to be the social bulwark, in the varied division of industries, in the widening opportunities to assert and maintain their individuality, women have escaped the pernicious condition which formerly darkened the best portions of Europe, under which, for want of occupation for independent maintenance, the daughters were shut into the alternative of an enforced marriage or an enforced convent.[4]

Unlike the Mount Holyoke graduates who listened to Bullock's speech, most telegraphers in the United States came from the traditional working class. Often their fathers were laborers or factory workers. Edwin Gabler, in *The American Telegrapher*, found that 65 percent of women telegraphers in the United States came from blue-collar backgrounds. Telegraphers' literacy, technical skills, and desire for respectability set them apart from their "rough working class" origins, to use the words of labor historian Stephen Meyer.[5]

Sarah Bagley went directly from the mills of Lowell, Massachusetts, into the telegraph office; as a highly literate member of the working class, she exemplified the profile of those who would follow her. Mary Stilwell, a telegrapher who became Thomas Edison's first wife, was the daughter of a wood sawyer. Many operators were the daughters of railroad workers.[6]

Although the spectrum of opportunities open to women telegraphers was significantly narrower than for their male counterparts and few women telegraphers could aspire to becoming industrialists or inventors, the occupation did create possibilities for social advancement. A few years' stint at the local depot might earn a promotion to a position in a larger city for a rural operator; study and persistence might lead to a position as a first-class operator or manager. And there was always the possibility, hinted at in numerous telegraphic romances, that the operator might meet and marry an eligible bachelor from the upper classes.

In England, women operators tended to come from a more middle-class background; according to Jeffrey Kieve, female telegraphers there tended to be the daughters of clergymen, tradesmen, and government clerks. Particularly after the telegraphic service came under the control of the Post Office, in the words of Charles Garland, "Women simply required a nomination for admission to the postal service, and their employment on a larger scale opened up a light and relatively remunerative occupation for the female relatives of postal officials."[7]

Many European administrations required that applicants for telegraphic positions pass an entrance examination; thus preference in hiring went to women with an advanced educational background. The Norwegian historian Gro Hagemann noted that when the telegraph service in Norway first admitted women in 1858, it recruited them from upper-class families because they were the only women who possessed the necessary intermediate school certificate. Thus telegraphy became a high-status occupation in Norway; Hagemann observed that "during the first decades of the existence of the telegraph, there were few, if any, professions open to women which enjoyed a higher status that that of the female telegraphers."[8]

In the United States, telegraphers believed their technical skills set them apart from members of the traditional working class. Their expressed opinions of other workers contain a note of elitism, perhaps an intimation of underlying anxieties about the narrow social gap that separated them from their own working-class origins. Nattie Rogers, fresh out of the country, in *Wired Love* found "the ignorance of people in regard to telegraphy" to be "surprising; aggravating, too, sometimes." A female telegrapher, interviewed by a reporter for the *New York World* during the 1883 strike, mentioned that she had visited the union hall but had not stayed long:

> "The fact is," she said, "there were not many ladies there, and the hall is not a pleasant one. It had been used by the striking cloakmakers and cigar-makers and it is dirty and disagreeable, and although most of the lady operators are willing to cooperate in every way to bring about the just demands of the telegraphers, they rather hang back from public demonstrations."
>
> "Why?"
>
> "Well, I suppose most of them are well-bred women with considerable refinement. They have to be to make good operators. I don't know any other reason. . . . You see a girl to do the work nice

has to have a fair education and to be pretty tolerably quick with her mind as well as her eyes, and that class of girls is a little better than some others."[9]

ETHNICITY

Most telegraphers in the United States in the early years tended to be native-born Americans with English-speaking ancestry, primarily because of the need for proficiency in written and spoken English. Later, immigrants of many nationalities began to see telegraphy as a means of gaining entry into the American middle class. A high percentage of telegraphers were Irish, probably both because of the relatively high degree of literacy of the Irish compared with other recent immigrants and the tendency of Irish women to marry later in life and thus spend more time in the workforce. Edwin Gabler found that of seventy women operators in New York City in 1880, 71 percent were of Irish ancestry.[10]

The profession was relatively free of ethnic discrimination, although the unions occasionally tried to exclude nonwhites and foreigners in the interests of job protection. When times were good economically and operators were scarce, telegraphers were inclined to democracy by permitting entry to the profession to all who possessed the necessary skills. At the 1865 convention of the National Telegraphic Union (NTU), the admission of women to the profession was debated, and it was noted without surprise that several African American men were working as telegraphers in the South. In the 1870s, the Cuban American brothers Ambrose and Narciso Gonzales began telegraphic careers in South Carolina that would eventually result in the founding of the Columbia, South Carolina, *State*, the newspaper they established to oppose the racist demagoguery of Benjamin R. "Pitchfork Ben" Tillman. The *Telegrapher* noted in 1873 that Ah-Bean, a Chinese man at the Sacramento, California, Western Union office, had learned enough Morse code to send and receive messages; at the same time, the Sacramento office of the Central Pacific Railroad employed a "young colored gentleman" who, while employed as a messenger, was also "somewhat proficient" in telegraphy.[11]

The census of 1890 provides some interesting information about the ethnic origins of telegraph operators, though it is necessary to filter some of

the racial assumptions built into the census report. The census differentiated 8,411 "white" female telegraph and telephone operators and 63 female
operators "of negro descent." (For male operators, there was an additional
category of "colored" persons who were not of Negro descent, which included Chinese, Japanese, and "civilized Indians.") Of the female operators
designated as white, 7,655 were identified as native-born Americans, and
756, or approximately 10 percent, were foreign born.[12]

Of the female telegraph and telephone operators, 3,599 were listed as
having foreign-born mothers. Of these, by far the largest number (1,907, or
53 percent) had Irish-born mothers; next were operators with German
mothers (650, or 18 percent), followed by England and Wales (458, or 13
percent), English-speaking Canada (225, or 6 percent), Scotland (145, or 4
percent), Scandinavia (69, or 2 percent), and France (31, or 1 percent). Less
than 1 percent of operators had mothers born in French-speaking Canada,
Italy, Hungary, and Bohemia in 1890.[13]

SCHOOLING

Since one of the primary requirements for telegraphy was a high level of literacy, most women who became telegraph operators had a formal public
school education. Almost all had a grammar school education, and many
were high school graduates. In the late nineteenth century, the number of
female high school graduates actually exceeded the number of male graduates; this pool of qualified women provided a source of labor not only for
clerical work, as Margery Davies noted in her 1982 study of female office
workers, *Woman's Place Is at the Typewriter,* but also for telegraphy.[14]

In addition to primary and secondary education, many telegraphers had
additional training, either at a business college that taught telegraphy or at
one of the telegraphic schools maintained by the telegraph companies.

The Cooper Institute

As part of its program to encourage the employment of women as telegraphers, Western Union sponsored a school in New York City to teach telegraphy to women. The concept of establishing a telegraphy course for
women was not original with Western Union; the American Telegraph

Company had already begun to train women operators before it was absorbed by Western Union in 1866. The school, known as the Cooper Institute, was jointly supported by Western Union and Cooper Union. Its course of instruction lasted almost five months and covered bookkeeping, battery maintenance, and report preparation, as well as Morse sending and receiving.

The *Journal of the Telegraph*, a Western Union–sponsored trade journal, described the purpose and goals of the school in its January 15, 1869, issue. Western Union's primary motivation for opening the school, according to the *Journal*, was economic: "With the pressure constantly being brought to bear on telegraph companies to cheapen rates, with competing lines drawing off that portion of the business which provided the margins of profit, and with the multiplication of short lines connecting factories and foundries with central offices, we expect to see demands made for women to serve in telegraph offices far beyond what now exists." The telegraph company rationalized hiring of women at lower rates of pay than men by claiming that the majority of women operators were not permanent members of the workforce but temporary workers who would leave when they married: "Marriage to them is home and an end of personal money-making for support. Thus they accept terms inferior to men because they need support only until the marriage state provides it." Western Union did, however, admit that some women might consider telegraphy to be a career and should be compensated accordingly: "There are, of course, exceptions. Women of executive ability command already salaries approximating those of men." The article closed by reassuring its largely male readership that the opening of the school posed no threat to their future employment or their privileged status in the operating room: "So far, man has proved the more reliable. He has an instinctive knowledge of business composition. This enables him to detect errors which a woman never suspects."[15]

The February 15 issue of the *Journal of the Telegraph* specified the rules and regulations for the Cooper Institute. The course of instruction was to last eighteen weeks, ending on July 1, 1869. Applications had to be written by the applicant herself, so that her literacy and handwriting skill could be evaluated. Potential applicants had to be over seventeen years of age but less than twenty-four and have good character references. Admission was highly selective; according to the *Journal*, "preference will be given to those who

by education and physical ability appear best qualified for the business."
Once a potential telegrapher was admitted to the institute, instruction was
provided without charge; however, regular attendance was mandatory, and
three absences without a "satisfactory excuse" were grounds for dismissal.
Pupils who were "wanting in diligence, or whose deportment is exceptional," were also likely to be dismissed. The hours of instruction were 9:30
A.M. to 3:30 P.M. daily. Pupils were not permitted to leave during the day,
and visitors were allowed, with special permission, only on Fridays.

Upon graduation, students were warned not to expect to be immediately placed in positions in the city: "Only pupils who are prepared to accept
situations out of the City of New York as soon as they are qualified, can be
admitted to the school." Applicants had to be prepared to accept a starting
position out of the city, perhaps in a rural depot; they would need several
years' experience to earn a coveted position in a large urban office.[16]

The results of this experiment in education were announced in the November 1, 1869, issue of the *Journal of the Telegraph*. Lizzie Snow, manager
of the city department of the New York Western Union office, was named
directress of the school; she was the role model that Western Union wanted
its trainees to emulate. It is clear that Western Union considered the school
to be a resounding success: "Three months ago the school was opened by
the admission of sixteen pupils. Of these two resigned, and four were found
incompetent. Of the remainder five have already been qualified to take regular situations, and three of these are already provided with places. A very
short time will serve to qualify the remainder for regular work, and it is apparent to all that the school is an unqualified success." The five graduates of
the first session were Isabella Sellew, Elizabeth O. Blanchard, Florence
Colyer, Armenia Frazee, and Fanny Oliver. The school was tough and competitive, judging from the dropout rate, and Lizzie Snow had already acquired a reputation as a hard taskmistress—a reputation that would
eventually prove her undoing. But for those who persevered, the rewards
were considerable: "The demand for these operators exceeds the supply,
and the remuneration is such that any young woman can support herself in
comfort, and, if she is very diligent and expert, could readily accumulate a
provision for that period when telegraphy will cease to be her occupation. . . .
This proves that practical telegraphy can be taught in a properly conducted
school."[17]

Peter Cooper himself attended a commencement of the school in late 1871; as the builder of one of the earliest locomotives in America and financier of the Atlantic telegraph cable, he took a special interest in education and had founded the Cooper Union in 1859 to provide free higher education to all. The October 16, 1871, issue of the *Journal of the Telegraph* gave statistics for the entire year; out of 275 applicants, 96 had been accepted as pupils; of these, 15 had resigned voluntarily and 4 had been dismissed. Of the 55 who had graduated, 40 had already obtained positions, and another 12 were "awaiting situations."[18]

The Cooper Union–Western Union telegraphic school for women would provide telegraphers to the industry for over twenty years, eventually graduating as many as eighty young women a year. Graduates began at branch offices; the more skilled might end up back in New York City as operators at the Western Union central office at 195 Broadway.

Armenia Frazee, one of the original 1869 graduates, was one of those who eventually came back to New York City to work at the main office. She was listed as an operator at the newly built main Western Union office at 195 Broadway in New York City in 1875, and nine years later she was still at the key there as a first-class operator.[19]

Telegraphic Colleges

Many business colleges began to teach telegraphy in the late 1860s and 1870s; in Cleveland in the late 1860s, a would-be telegrapher could choose between the Cleveland Business and Telegraphic College and the Cleveland branch of the Bryant and Stratton Business and Telegraphic College. Bryant and Stratton was a chain, with schools in thirty-three cities.

Telegraphers in general had a low opinion of these telegraphic colleges, maintaining that the only proper place to learn telegraphy was at the elbow of an experienced operator. The editor of the *Telegrapher*, speaking of Bryant and Stratton, opined in 1865 that "to teach telegraphing in one of these colleges is like teaching a boy to swim on dry-land; the element is as much lacking in one as the other."[20]

Women who attended these colleges had to deal not only with the general prejudice against hiring women but also with the prejudice against hiring graduates of telegraphic colleges. Nettie Bronson wrote to the *Teleg-*

rapher in 1873 to describe the difficulties she encountered in trying to find a position after graduating from a telegraphic college:

> I learned telegraphing in an institute, and after having been there about seven months, I tried to get into a commercial office, knowing that I could learn faster, and at the same time learn how the business was done. The answer that I received was, "The Superintendent does not allow students." The respondent did not care a fig for what the Superintendent said. If it had been a boy he might have taken him, but you see I was a girl (a fault in my father's education). The reason was he could not sit in a chair with his feet on top of the stove and throw tobacco spit the whole length of the room. He could not sit and smoke a cigar without saying (if he was a gentleman) "please allow me to smoke," or "is it offensive to you?"[21]

Alternatively, a woman could respond to newspaper advertisements offering private instruction, usually by moonlighting operators. Such an advertisement appeared in the *Cleveland Leader* in 1874: "Wanted—A few ladies to learn telegraphing; inquire at 37 Howard st. in the Heights. Tuition paid monthly. The quicker learned the less expense."[22] Many of these ads turned out to be fraudulent. The April 15, 1882, issue of the *Operator* carried the story of a group of women who went to the police station in Chicago to demand a warrant for the arrest of E. G. Chapman, a "sleek-looking fellow, with silky black side whiskers," who had promised to teach them telegraphy and find them positions, at a rate of $15 to $50 each. According to the women, he had "never taught them anything."[23]

To supplement their formal telegraphic education, many telegraphers purchased a practice set, consisting of a telegraph key and buzzer, so they could practice at home in their spare time. It was time well invested because speed and accuracy were the primary requirements for raises and promotions. For Ma Kiley, setting up a practice set at her parents' home in Del Rio, Texas, in the late 1890s was the first step to a career at the key: "Mother let me put my key on her breadboard and I printed the code beside it. . . . I helped mother with the work and put in every minute I could on my practice. I loved it. When I read the papers or a book or anything I could get my hands on, I'd be sending what I was reading."[24]

The job itself provided an excellent education. Telegraphers were in daily contact with the news of the world, business transactions, commodity prices, and the ordinary social activities of the community. Women telegraphers in small towns were probably among the best-informed persons in their communities.

Mentoring

Mentoring was an important element in the careers of many women telegraphers. Acquiring the skills necessary to become an operator necessitated spending some time at the elbow of an experienced operator; it was the only way to gain a "hands-on" appreciation of what was involved in the work. Elizabeth Cogley owed her entry into railroad telegraphy in the 1850s to the fact that the previous Lewistown, Pennsylvania, operator, Charles Spottswood, boarded with her family and taught her the basics of the craft. Nearly fifty years later, Ma Kiley would acquire telegraphic skills from Henry Hall, who stayed at the hotel run by her parents in Del Rio, Texas, around 1900.[25]

In some cases, the mentor was a woman. The women operators of the Deseret Telegraph in Utah often passed telegraphic skills from one generation to the next. The Utah historian Kate Carter tells the story of the Johnson family in her book on the operators of the Deseret Telegraph, *The Story of Telegraphy*. Mary Ann Johnson was the fourth wife in the plural marriage of Aaron Johnson, one of the original Mormon settlers of Springville, Utah. After the completion of the Deseret Telegraph in the late 1860s, she was one of the first to volunteer to learn telegraphy and serve her community as operator. After learning telegraphy herself, she taught the craft to her daughter Ina. Ina was put in charge of the telegraph office in Lehi, Utah; she in turn taught telegraphy to her daughters Celestia and Tina and her niece Ada. Ina also taught telegraphy to other women in the community, including Barbara Evans, Isabella Karren, and Harriet Zimmerman. Celestia and Tina carried on the family tradition, working as career railroad operators with the Oregon Short Line. Tina also taught the profession to her brothers Bruce and Mitchell.[26]

Mentoring among women operators was not confined to the Mormon operators of the Deseret Telegraph. Mary Macaulay, a career press operator and official of the Commercial Telegraphers' Union of America, learned telegraphy in 1878 at the age of thirteen at the railroad depot in Leroy, New York, from Nellie Chaddock, the depot operator.[27]

Training Teletype Operators

When the Teletype came into general use around 1915, Western Union set up a school to teach Teletype operation in New York City. Located two floors above the operating department at 24 Walker Street, the school taught touch typing and printer operation, as well as geography, spelling, and copy-reading. These courses, together with a six-month apprenticeship, were required in order to become a printer operator.

Western Union later opened similar schools to teach Teletype operation in other parts of the country, including one in Crawfordsville, Indiana. Like many other women from small midwestern towns, Virginia Brom took the local train, or "doodlebug," from Pella, Iowa, to Crawfordsville to attend the school in 1944. After completing the course, she worked as an operator in Pella, Ames, and Carrolton, Iowa, during World War II, when many women entered the occupation because of the shortage of male operators.[28]

Training in Other Parts of the World

Training in England in the late nineteenth century was similar to that offered by the Cooper Union school but of shorter duration. Applicants to the Electric Telegraph Company's London school on Telegraph Street in Moorgate had to be between fourteen and eighteen years old. New employees of the Electric Telegraph Company were trained without pay for six weeks. If they did not reach the required standard of eight words per minute by the end of this period, they were dismissed.[29]

In 1877, women who wished to be admitted to the telegraphic service in Paris at the Central télégraphique in the Rue de Grenelle had to pass an entrance exam; subjects covered included writing, penmanship, arithmetic, the metric system, and geography. Trainees between the ages of sixteen and twenty-five were accepted; they had to be single or widows without children. Those who passed the exam received a starting salary of eight hundred francs per year, with annual increases to about fifteen hundred francs per year.[30]

In many European administrations, it was necessary to pass examinations in order to qualify for advanced positions. In Italy around 1900, two women were selected from each office to serve as *assistants*, or supervisors. Women who wished to qualify for these positions had to pass an examination covering the Italian and French languages, arithmetic, physical geography, physics, chemistry, political geography, the theory and practice of telegraphy, and the regulations of the telegraphic service. Women who passed this exam were considered to be permanent employees of the telegraphic service and were granted permission to remain in the service after they married; nonsupervisory female telegraphers were required to resign their positions if they married.[31]

The Chilean National Institute began offering telegraphic training to women sometime before 1870. An applicant who completed the course at the institute would be preferentially selected to fill vacancies, although applicants with practical experience in a state telegraph office were also considered. Applicants were required to be eighteen years of age or older and to have passed an examination in the use of telegraph instruments. Successful applicants were required to serve a two-month apprenticeship, after which they received a certificate of employment.[32]

REASONS FOR ENTERING THE WORKFORCE

Many women who became telegraphers in the nineteenth century did not follow the standard pattern of being supported first by a father and later by a husband. They often came from a family with an absent or irregularly employed father. Most worked for only a few years and then left to marry and raise a family; some, however, regarded telegraphy as a career and remained in the profession for many years. Some contributed to the support of their family, like Emma Hunter, who was helping to support a widowed mother and brother; others, like Fannie Wheeler, simply wished to live an independent life.

For many women, telegraphy was attractive work. It was clean work and socially above factory or domestic work; in the words of Martha Rayne, author of *What Can a Woman Do?* published in 1893,

> It does not soil their dresses; it does not keep them in a standing posture; it does not, they say, compromise them socially. A telegraph operator, they declare, has a social position not inferior to that of a teacher or governess. . . . Moreover, the young women operators at the Western Union Company's headquarters are treated by their superintendent—a young woman very proficient in her profession—with sedulous courtesy. She addresses them, not familiarly by their Christian names, but by their surnames, with the prefix "Miss," and she insists upon their addressing one another in the same considerate fashion. . . . When a message has been dispatched or received, the operator may, and often does, take up her knitting, crocheting, or sewing, passing pleasantly the interval until the arrival of the next message.[33]

Having to stand all day without being able to sit down was considered dangerous for young girls in the Victorian age; it was believed to lead to sterility. Thus the ability to sit while working was especially reassuring for those who intended to marry and have a family.

In England as well, telegraphy was looked upon as desirable work; on one occasion in 1862, according to Jeffrey Kieve, an advertisement for telegraphers in a morning newspaper brought out four hundred "respectably dressed females," prompting another paper to question what societal dysfunction was responsible for "so large a number of respectable and well-to-do young women" applying for employment as telegraphers.[34]

In other parts of the world, women sometimes were encouraged to become telegraphers as part of state policy. In Sweden, the States-General reported to the king in 1863 that a considerable surplus of women over men of marriageable age existed, and "consequently a large number of women were cut off from their natural vocation, marriage." The States-General, feeling that it was the responsibility of government to find work for these women, recommended that women be allowed to work as telegraphers; the

Swedish government agreed and directed the state Telegraph Administration to admit women to telegraphic service. Similar actions were taken in Prussia and Australia in the nineteenth century.[35]

DEMOGRAPHIC COMPOSITION OF THE WORKFORCE

The number of women telegraphers listed in the U.S. census ranged from 355 in the 1870 census to almost 17,000 in 1920, the peak year for the number of telegraphers reported by the census. Thus the number of women employed as telegraphers was far less than those employed in more traditional occupations for women, such as domestics or teachers, but greater than the number of female attorneys. Female physicians outnumbered telegraphers in 1870 but were outnumbered by telegraphers by 1920. Table 1 compares the numbers of women employed in the United States in various occupations for the years 1870–1920.

U.S. Census Figures for Telegraphers Based on Region, 1870–1960

The United States census of 1870, the first to differentiate occupations by gender, showed that 4 percent of telegraphers were women. By the 1920 census, 20 percent of all operators were female. Among teletype operators, a majority were women after 1930. Table 2 shows the number of women employed as telegraphers in different regions of the United States for the years 1870–1960.[36]

The Validity of the Census

Reading the popular literature and trade journals of the late nineteenth century gives the modern reader the impression that women telegraphers were more numerous than the census figures suggest. Census figures in general in the nineteenth century were of dubious accuracy, particularly regarding the employment of women. In her 1992 article "The History of Women and the History of Statistics," Margo Anderson notes that the census of 1870 was able to record occupations for less than half of the 28.5 million people listed as being over the age of ten; for women, only 16 percent were listed as being gainfully employed, far lower than one would expect. More than eight million women were simply missing from the census rolls.[37]

Anecdotal accounts from the late nineteenth century often show much higher percentages of women employed as telegraphers than the census figures indicate. Although Eugene J. O'Connor, chairman of the Executive Board of the Brotherhood of Telegraphers, and Walter C. Humstone, superintendent of Western Union, agreed on little else during the strike of 1883, they both cited figures for the employment of women as telegraphers that were far in excess of those reported by the census. O'Connor, testifying before the Senate Subcommittee on Labor and Education in August 1883, estimated that 20 percent of all telegraphers were women; Humstone, speaking to *New York Herald* reporters in July of the same year, stated that

TABLE 1
WOMEN IN VARIOUS OCCUPATIONS IN THE UNITED STATES, 1870–1920

Occupation	1870	1880	1890	1910	1920
Domestic servants	867,354	938,910	915,927	1,309,549	1,012,133
Teachers	84,047	154,375	237,508	478,027	639,241
Telegraphers	355	1,131	8,474	8,219	16,860
Physicians	525	2,432	4,557	9,015	7,219
Lawyers	5	75	208	558	1,738

TABLE 2
WOMEN TELEGRAPH OPERATORS IN THE UNITED STATES, 1870–1960

Year	Northeast/ Mid-Atlantic		South		Midwest		Great Plains/ Far West		Total U.S.	
	M	F	M	F	M	F	M	F	M	F
1870	4,510	277	647	1	2,115	45	689	32	7,961	355
1880	11,217	785	1,765	43	6,494	236	2,202	67	21,678	1,131
1890	18,791	5,089	4,704	442	13,192	1,912	7,053	1,031	43,740	8,474
1900	20,621	3,794	6,127	532	14,761	1,885	7,134	1,018	48,623	7,229
1910	23,919	3,643	8,522	752	16,974	2,084	12,319	1,740	61,734	8,219
1920	23,815	6,312	8,525	1,950	16,315	4,291	13,419	4,307	62,574	16,860
1930									51,699	16,122
1940									31,554	8,228
1950									27,090	7,290
1960									15,980	4,496

"about twenty-five percent of the operators are women." The census of 1880, however, showed only 5 percent of telegraph operators to be women.[38]

Railroad Operators

Statistics on the number of railroad telegraphers in the nineteenth century are notoriously unreliable, not only because of the inaccuracy of most census reporting but also because of the work habits of railroad operators, "boomers," who changed jobs frequently. Archibald M. McIsaac, in his 1933 study of railroad telegraphers, guessed that about two-thirds of the telegraph operators in the United States in 1883 were railroad operators. Since the census of 1880 recorded about 22,000 telegraph operators, this would indicate around 14,600 railroad operators in the early 1880s. Likewise, the census reported about 5 percent of all telegraphers to be female in 1880; this would lead to a figure of about 730 female railroad telegraphers in 1880. (These figures almost certainly err on the low side, both numerically and in percentages.) The problem of accurate reporting is compounded by the fact that railroad operators did not begin to think of themselves as being occupationally separate from commercial operators until the late 1880s, when they formed their own labor organization, the Order of Railway Telegraphers (ORT).[39]

In the early twentieth century, figures became more precise as census figures improved in accuracy and occupations were more clearly differentiated. U.S. Bureau of Labor statistics show the number of women employed as railroad telegraphers continuing to climb in the late nineteenth and early twentieth centuries, peaking in the early 1920s. Of the 78,000 railroad telegraphers employed in the United States and Canada in 1920, 2,500—a little over 3 percent—were women.[40]

Regional Demographics in the United States

In the United States, the largest numbers of female operators were found in the Northeast and Middle Atlantic states in the mid-nineteenth century, reflecting not only the early development of the telegraph industry in this area but also the greater maturity of industry and commerce. As the telegraph lines began to move southward and westward later in the century as those regions began to develop, the demographics began to shift accordingly. The census figures show 78 percent of all female telegraph operators to be employed in the Northeast and Middle Atlantic regions in 1870; by 1900 this figure had fallen to 52 percent, and by 1920, only 37 percent of all women telegraphers worked in the Northeast and Middle Atlantic states.

Women Operators in the South. — Morse's original lines were extended southward from Washington in the late 1840s and early 1850s, passing through Richmond, Virginia, Raleigh, North Carolina, Charleston, South Carolina, Savannah, Georgia, and Mobile, Alabama, finally terminating in New Orleans, Louisiana. As in the North, the employment of women in telegraph offices probably increased during the Civil War as men were drafted or enlisted in the army.

More women began to be employed as telegraphers after the Civil War as white southern women entered the workforce in large numbers, both because of the depressed economic condition of the postwar South and the loss of male breadwinners. Some southern men found this change difficult to accept; as Robert M. Cowan, editor of the Wadesboro, North Carolina, *Anson Times* remarked in 1884, "Of the multitudinous changes brought about by the result of the war, there is none so distressing to us as the necessity which forces some of our true and brave hearted women to earn their living by contact with the world." He then proceeded to enumerate the types of employment that women were finding in the postbellum South: "They were given clerkships, made amanuenses and copyist, taught telegraphy, phonography, and type letter writing, and today can be seen filling positions formerly occupied by robust men." In closing, though, Cowan showed his progressive views; he encouraged his readers to support women in their struggle to earn an independent living, especially in small towns like Wadesboro: "There are but few openings for a woman in a small place to earn a livelyhood, and they should be encouraged in their labors as clerks. It is noble for them to pocket their pride and do battle with the stearner [*sic*] sex in this scramble for a living, and they should be encouraged, not only by a smile of approval but by that potent and powerful influence, patronage, which brings with it a feeling of independence and satisfaction."[41]

As in other parts of the country in the post–Civil War era, women who regarded telegraphy as a career moved from town to town in the rural

South. Alice F. Johnston, a native of Maryland, telegraphed in Aiken, South Carolina, before moving to Wadesboro, North Carolina, in 1882 to work for Western Union. The *Anson Times* took notice of her arrival and her apparent ability to turn heads:

> The Western Union Telegraph Company, having transferred our friend Jas. Eason to another field, have secured the services of Miss Alice Johnson [*sic*], recently of Aiken, S.C. Miss Johnson arrived Saturday and took charge Monday. We trust she will find her new position a pleasant and desireable one, and her stay among us be both long and agreeable. Miss J. understands her business well—and we are inclined to suspect that there will be more cotton telegrams sent than formerly by some of our younger dealers.[42]

Telegraph schools for women began to open in the South in the late nineteenth century. Ola Delight Smith learned Morse code on a practice set at home in Epes, Alabama, around the turn of the century and then attended a telegraphy course at the Alabama Polytechnic Institute for Girls in Montevallo before going to work for the Queen and Crescent Railroad in 1900.[43]

Women Operators in the West. — The expansion of the telegraph system into the prairie region occurred in the late 1840s and early 1850s, as telegraph lines moved westward from Cleveland, Ohio, to Detroit, Michigan, and onward to Chicago, roughly following the southern margin of the Great Lakes. The lines were also extended southward across Illinois, reaching St. Louis at about the same time. Westward expansion across the plains occurred in 1861, when a consortium of telegraph companies completed the transcontinental telegraph, which stretched from Omaha, Nebraska, to San Francisco, California, passing through Nebraska, Wyoming, Utah, and Nevada.

Because there were few large metropolitan areas in the West, the majority of positions available to women were in rural railroad depots. *Electrical World* noted in 1886: "Far out on the western plains, wherever there is a road station, almost invariably the traveler sees a pretty lace or muslin curtain at the window," a sure sign of the presence of a female telegraph operator within.[44]

The arrival of the transcontinental telegraph in Salt Lake City in 1861 stirred interest in the Mormon community in creating a telegraphic system to connect the widely separated Mormon settlements in Utah, Nevada, and Idaho. The leader of the Mormons, Brigham Young, had participated actively in the construction of the transcontinental line, supplying timber for telegraph poles and work crews to help set the poles; evidently as a reward for his cooperation, the telegraph companies decided to use Salt Lake City as the location for the joining of the wires from the East and the West. As a result of their participation in building the transcontinental line, Mormon leaders began to realize the value of a telegraph system in unifying the isolated Mormon settlements scattered across the West.

Construction on the Deseret Telegraph, as the project came to be called, began almost immediately. Poles were set and wires were strung to each settlement from the central office in Salt Lake City. Construction of the Deseret Telegraph was completed in 1866. Each community then selected young men and women to serve as telegraphers and sent them to Salt Lake City for instruction.

Women were employed as operators from the beginning, and they formed a considerable proportion of the operators. Although the Mormon community was organized along patriarchal lines and generally gave women a secondary role, the notion of "separate spheres," as understood in much of nineteenth-century America, was subordinated to the ideal of community service within the Mormon community, and women were allowed to serve in a variety of nontraditional roles. Thus telegraphy became a "family tradition," with as many as three generations of women working as operators.[45]

Women operators in the West seemed to have encountered less gender-based discrimination than their eastern counterparts. This was probably owing in part to both the more relaxed social codes of the West, which acknowledged that women might have to enter nontraditional roles, and the general scarcity of women on the frontier; women wage earners were not seen as a threat to male jobholders, as in the East.

The percentage of telegraphers who were women was slightly lower in the West than in the East. The census of 1870, the first to differentiate occupations by sex, showed 4 percent of telegraphers to be women

Figure 18. Emily Warburton and Barbara Gowans, Deseret Telegraph, Tooele, Utah, 1871. From Carter, "The Story of Telegraphy," in *Our Pioneer Heritage*, 561. Courtesy Daughters of Utah Pioneers.

nationwide, while less than 3 percent were women in the Midwestern/ Prairies and Far West/Plains regions combined (see Table 2). By 1900, when the census showed 13 percent of operators nationwide to be women, the West still trailed the East with only 12 percent of woman operators. Reflecting the general westward population shift, however, the proportion of women operators employed in the western states grew from around 20 percent of all women operators in 1870 to 40 percent in 1900.

Demographics in Other Parts of the World

As in the United States, the highest percentages of women telegraphers in Canada tended to work in the large telegraph offices in the eastern part of the country. Shirley Tillotson found that in 1902, 42 percent of the operators at the Toronto office of the Great Northwestern Telegraph Company were women, as were 28 percent of the operators at the offices of the Canadian Pacific Railway (CPR) Telegraph in the same city. A smaller percentage of women worked in offices in western Canada, particularly before 1900; only 5 percent of the operators in Winnipeg were women in 1894, but the figure was 18 percent by 1917.[46]

The percentage of female operators in England and Europe during the late nineteenth century seems to have been somewhat higher than in the United States, at least in urban areas. In England, women were employed both in the large commercial telegraph offices in London and in smaller offices in towns and rural areas. In 1870, shortly after the telegraphic service came under the control of the British Post Office, 1,535 of the total of 4,913 telegraphers, or 31 percent, were women.[47]

In France, women worked both at large offices in the major cities and in small rural offices. In the Paris Central télégraphique in 1880, 230 of the 624 operators, or 37 percent, were women in 1880; two years later, the number of female operators had increased to 300, or nearly half of the total.[48]

As in the United States, the percentage of women employed as telegraphers in Europe continued to rise in the early twentieth century. In Austria in 1930, 2,735 of the 5,900 employees of the telegraphic administration were women, or 48 percent; this figure includes all telegraphic administration employees, from housecleaning personnel to senior administrators. Of the operators, 250, or 47 percent, were women, but only 2 were in managerial positions.[49]

Age of Women in the Telegraphic Workforce

When Nattie Rogers asks her beau Clem for his age over the line in Ella Cheever Thayer's 1879 novel, *Wired Love*, he replies "Did you ever see an aged operator? I never did, and don't know whether its because electricity acts as a sort of antidote, or whether they grow wise as they grow old, and leave the business." Telegraphy in the mid-nineteenth century was a young person's game, and women operators in particular tended to be young and single. The average age of 102 women operators in New York City in 1880 was 21.8 years. In the 1880s, however, women of all ages could be found working in the telegraph office, from 11-year-old Nellie Welch, operator of the Point Arena, California, telegraph office in 1886, to Elizabeth Cogley in Harrisburg, Pennsylvania, who at the age of 55 in 1889 was one of the most senior telegraph operators in the United States. Because of the relative youth of the occupation, few women older than Elizabeth Cogley worked in telegraph offices in the 1880s; later, many women worked until retirement age—and beyond. After nearly fifty years of service in the telegraph office in the United States and Mexico, Abbie Struble Vaughan retired at the age of 67 in 1912—only to come out of retirement five years later, at the age of 72, to teach telegraphy during World War I.[50]

It was not unusual for either boys or girls to enter the profession at the age of thirteen or fourteen. At a time when child labor laws were either nonexistent or weakly enforced, an adolescent who demonstrated persistence and aptitude could volunteer to substitute on an informal basis for the regular operator and eventually acquire a paying position. Mary Macaulay, who later became a career press operator and the vice-president of the Commercial Telegraphers' Union of America, started out in the depot at Leroy, New York, at the age of thirteen in 1878. Similarly, Medora Olive Newell learned telegraphy at the depot in Durango, Iowa, at the age of fourteen, her first stepping-stone to a career as a dispatcher and first-class operator. Starting out at an early age seems to have been particularly common among women who later became managers or career first-class operators.[51]

The majority of women tended to leave the workforce to marry after a few years. Of the 196 women who left the telegraphic service to marry in

England in 1899, the average age was twenty-seven, and the average number of years spent as a telegrapher was between eight and nine. In Belgium at about the same time, the average length of service was approximately five and one-half years.[52]

Salaries in the United States

In the beginning, salaries were set arbitrarily because it was not clear what the economic impact of the telegraph would be, and the earliest telegraph companies were perennially strapped for cash as they simultaneously attempted to extend their lines and operate at a profit. Emma Hunter was initially offered $50 a year in 1851 for her services at the West Chester, Pennsylvania, office of the Atlantic and Ohio Telegraph Company; her contemporary, Helen Plummer, made $125 a year to run the Greenville, Pennsylvania, office. Virginia Penny believed that most women operators made between $6 and $25 a month by 1860.[53]

Salaries increased sharply during the Civil War because of the shortage of operators but declined after the war's end; thereafter they tended to remain fairly constant, at least in numerical terms, from the late 1860s through the early years of the twentieth century, although buying power fluctuated fairly widely. Generally, a manager would earn $100–150 a month (though managers in small towns often made no more than operators in a big city); a chief operator, $75–100; first-class operators, $60–100; second-class or beginning operators, $15–60; clerks, $15–30; and messenger boys, $10–15 a month. Pay was higher in the large cities and lower in rural areas. Pay was determined on an individual basis; while most women operators made less than $50 per month, "superstars" like Ma Kiley and Mazie Lee Cook could make as much as $150 or $160 a month around 1910. A position as an operator at a resort or hotel often included free room and board. These jobs offered low pay—typically $15 or $20 per month— but offered opportunity for independence, adventure, or just a pleasant summer job. Some of the highest-paying jobs for women operators were at the brokerage houses, where an expert nonmanagerial operator might expect to make as much as $90 a month.[54]

The ledger for the Harrisburg, Pennsylvania, Western Union office, now part of the Western Union Collection located at the Smithsonian Archives, gives payrolls for that office covering the period 1861–79. Salaries

for the month of May 1867 are given in table 3; they are probably representative of the salaries paid in many medium-sized telegraph offices across the
United States during the post–Civil War era.

Presumably, all are male with the exception of those designated "Mrs."
Mrs. Lunger replaced a Miss Springer who had appeared on the payroll the
previous month; it is possible that they are the same person. Clearly Mrs.
Lunger did not make as much as Ziegler or Tyndall, presumably first-class
operators, but she made considerably more than Wilson and Kiefer, who
were probably novice beginners. Mrs. Sanborn was a fairly highly paid
clerk; perhaps she had acquired telegraphy skills and could function as an
operator when required.[55]

Beginning female telegraphers made \$15–45 per month. Starting pay
was about the same for men and women, but men moved up more rapidly to
a maximum second-class salary of about \$60 a month. Wage discrimination
existed for first-class operators as well; while a male first-class operator
might make \$80–85 in the 1880s, his female counterpart might make only
\$75 for the same work.

Relationship to Other Types of Women's Work

During the period 1860–90, the average pay of a woman telegrapher was
approximately equal to that of a female teacher or bookkeeper in the United

TABLE 3
PAYROLL FOR HARRISBURG, PENNSYLVANIA,
WESTERN UNION OFFICE—MAY 1867

Name	Title	Salary
W. D. Sargent	Manager	110.00
R. P. B. Ziegler	Operator	75.00
J. B. Tyndall	"	65.00
Mrs. Lunger	"	30.00
W. G. Wilson	"	20.00
A. R. Kiefer	"	16.66
Mrs. Sanborn	Clerk	30.00
W. S. Rupp	Messenger	12.00
Geo. Weitzmann	"	12.00
Mrs. D. Reed	Janitress	10.00

States. It was far above the $10 or $15 a month that a female factory worker earned or the $5 or $10 a month that a domestic might earn. Telegraphers were about on a par with nurses in 1860 but fell below them by 1890, when telegraphers' pay fell and nurses' pay rose. William Shanks listed the following weekly wages as typical for women in 1868: bookbinder, $10; seamstress, $4.50; telegraph operator, $10; schoolteacher, $12; and actress, $18.[56]

The rise of the telephone industry led to a new occupation, that of telephone operator. It would be wrong, however, to assume that telegraphers "evolved" into telephone operators; the work differed greatly. Being a telephone operator required a less technical skill set, focused primarily on switchboard operation; telephone operators were trained to connect telephone users quickly by means of patch cords, based on verbal instructions and visual status indicators, under heavy supervision. Telephone operation quickly became a gendered occupation; by 1900, most telephone operators were female.

During the late nineteenth century, the wages of telegraphers were far above those of telephone operators; by the 1920s, as the wages of telephone operators increased and Morse operators were replaced by lower-paid Teletype operators, the differential narrowed. The *Census of Electrical Industries* gave average annual salaries of $544 for telegraphers and $408 for telephone operators in 1902; the same source showed a much narrower differential, $1,110 for telegraphers and $1,064 for telephone operators, in 1922. By 1944, telegraphers' union figures showed that telegraphers on average earned less than telephone operators—$37 a week for telegraphers and $39 a week for telephone operators—a measure, at least in the eyes of the union officials, of how far the state of the telegraphers had fallen in economic prestige.[57]

In general, women telegraphers were paid less than men for the same work. It appears that women in the United States were paid two-thirds to three-quarters of what a man would make for the same work. After the Civil War, during the economically slow early 1870s, Western Union made it a policy to hire new employees at a rate slightly less than the employee replaced—a practice referred to as the "sliding scale." It also actively encouraged the hiring of women. Many novice women telegraphers were hired to replace more experienced male telegraphers at lower pay rates. It thus is difficult to separate out experience from sexism, though both were certainly factors.[58]

Many women left the field after only a few years. Those who remained and became career managers or operators often made salaries close to those commanded by their male counterparts. Nevertheless, "equal pay for equal work" was an important factor in several of the strikes and labor disputes in the United States.

A single woman could survive on the pay she made as a telegrapher, but she had little opportunity to save money, especially at entry-level wages. Women telegraphers were expected to dress for their profession; as Martha Rayne commented, a fair proportion of a beginning telegrapher's income went for her work dresses, hats, and shoes:

> Here is a young woman, say eighteen years old, in the second year of her course [i.e., two years of actual experience]. Her pay, we will say, is as yet only thirty-five dollars a month, and if she depends entirely upon her earnings for support, she is likely neither to save a cent nor to make a cent. Her board and room will cost her probably at least six dollars a week, or if she has a room-mate, possibly five dollars; her luncheons, her car fares, her washing, half as much more, without any extravagance on her part; her office dress, even if she make it herself, will take eight dollars out of her pocket-book; her bills for other clothes, for shoes, for hats—well, it is easy enough for her to expend ten dollars every week in the year, and her salary is not nine dollars.

Martha Rayne paints a grim picture of the earning potential of a woman entering telegraphy in 1893; in fact, she discourages women from considering telegraphy as an occupation. But she also notes that a few years earlier, wages had been higher and expenses lower; in 1893, a year of panic and depression, the industry was overcrowded, jobs were scarce, and wages low. A large percentage of beginning telegraphers lived at home, enabling them to save some of their income. Rayne is a bit vague about clothing expenses; a study done in 1915 showed that dressing for this type of work would cost around $120 a year—three or four months' pay for a beginner.[59]

While the gender gap in pay was certainly real for Morse operators, it was also rather nebulous and greatly affected by factors such as local conditions, the skill of the operator, and her persistence as a negotiator. As the work became increasingly gendered with the introduction of the Teletype, the gender inequity became increasingly institutionalized. When the Commercial Telegraphers' Union signed a contract with the United Press in 1937, it agreed to a wage range of $52.50 to $55.00 per week for the predominantly male Morse operators and $45.00 to $47.50 a week for the predominantly female automatic operators.[60]

Salaries in Other Parts of the World

Salaries for women operators in Canada who worked for railroads and private telegraph companies were similar to those in the United States. In the maritime provinces, however, the Government Telegraph Service paid female telegraph agents $50 per year between 1880 and the 1920s, a below-subsistence wage which the operator's families were expected to supplement by gardening, fishing, and other occupations.[61]

In England, a beginning female telegrapher could expect to earn around 10 shillings a week in the 1860s and 1870s, equivalent to about $2.50 U.S. An experienced operator capable of sending and receiving twenty-seven words per minute (roughly equivalent to a first-class operator) could expect to make about 30 shillings a week, or about $7.50 U.S. The average rate of pay for women operators rose steadily during the Victorian age, from around 17 shillings in 1872 to 26 shillings in 1897.[62]

In France in 1880, women telegraphers at the Central télégraphique made between 1,000 and 2,000 francs per year (approximately $140 to $280). This was almost exactly half the pay male telegraphers received at the same time.[63]

Women operators in India in the early part of the twentieth century started at an annual salary of 40 rupees; after four years on the job, they were eligible for an annual increment of 2 rupees, 8 annas, up to a maximum of 80 rupees. In Rangoon, Burma, at the same time, the pay range for women telegraphers was 50–90 rupees, or about half the pay given to male telegraphers, who could earn as much as 150 rupees per year.[64]

SOCIAL LIFE

Some social perks came with the work; both the telegraph companies and the labor organizations held picnics and parties. Telegraphers were enormously fond of parties, especially dances; the New York District Telegraphers' Ball, held on April 22, 1865, by the members of the National Telegraphic Union, lasted all night, with the last quadrille being danced at 5 A.M. The hall had been professionally decorated, with red, white, and blue bunting and a bronze medallion of Samuel Morse, by a "celebrated decorator" named Maximillian. Between three and four hundred tickets were sold, and a profit of $270 went into the treasury of the NTU. The editor of the *Telegrapher* noted the presence of at least five women telegraphers—Mrs. Lewis and Misses Snow (probably Lizzie Snow), Avery, Hinds, and Turner, all employees of the American Telegraph Company; he mentioned that at least one of the ladies danced every set until half past four.

Balls and dances formed an important part of the social life of telegraphers, who rarely had time to socialize during the working day and were discouraged from doing so by management. For women in particular, the dances provided an opportunity to socialize, sometimes all night long, in a respectable and chaperoned environment.

Excursions were sometimes provided free of charge by the railroad companies as a token of appreciation for the services of the telegraphers, upon whom the railroads depended heavily. One such outing occurred on February 19, 1876, when the Central Railroad of New Jersey treated the telegraphers of New York City to a train trip to the beach at Long Branch, New Jersey. Telegraphers who showed up on time at 11 A.M. at the ferry house at the foot of Liberty Street in New York were taken by boat to Jersey City, where they boarded a special excursion train, with live music provided by a band.

Forty-three operators and spouses sang along with the band as they traveled; there were eight married couples, nine single women, two married women, each with a child, and fourteen single men. Among those present were Frank Pope and J. N. Ashley, editors of the *Telegrapher*, and Frances L. Dailey, manager of the city department at Western Union. The train stopped at Squan, where the telegraphers received a free dinner at the local hotel, and then proceeded on to Long Branch, where a dance was held at the railroad depot; some walked to the beach nearby. At 4:15 P.M., the train started back toward Jersey City, carrying a load of tired but happy revelers; as Pope and Ashley reported, "To telegraphers, whose business confines them so closely to their offices and to arduous labor, such an occasion is a grateful relief, and it is to be hoped that they may occur more frequently hereafter." The editors' hopes were realized in July of the same year, when the steamship *Fort Lee* set out from New York City's Twenty-fourth Street pier for a voyage up the Hudson, carrying participants in the "First Annual Excursion and Picnic of the New York Telegraphers Association" to a dance at Pleasant Valley. The usual waltzes, quadrilles, and schottisches were danced, including Strauss's "Telegramme" waltz. The ever-vigilant editors of the *Telegrapher* hinted broadly at the romantic possibilities of such an excursion, noting that "there were numberless shady nooks and quiet walks which the gentlemen discovered, in company with—well, they did not go alone, we assure you."[65]

Telegraphers enjoyed the arts, and not just as spectators; they put on numerous theatrical and musical productions. The New York Telegraph Operators' Dramatic Club staged a presentation of Mark Twain's *Tom Sawyer* in 1893, featuring May Saunders and Marion McLaren, both telegraphers, in starring roles; the entertainment also included singing by Kitty Stephenson, also a telegrapher.

In the 1890s, when bicycles became a popular fad, women telegraphers became enthusiastic cyclists. Mrs. S. E. Sandberg and Mrs. L. C. White, already known to readers of *Telegraph Age* as speed sending and receiving champions, joined a New York bicycle touring club in 1893. Jennie Chase, manager of the Postal Telegraph office in Oshkosh, Wisconsin, in 1894, was cited as having "the distinction of introducing bloomers, the popular bicycle costume, in that city."[66]

TRAVEL

One of the distinguishing characteristics of women telegraph operators in the nineteenth century was a relatively high degree of mobility. Railroads often offered free passes to telegraphers, enabling them to move easily from place to place in search of work. Moving around was a way of advancing one's pay scale relatively quickly; staying in one place and waiting for a promotion often took years, whereas relocating to a distant city to fill a vacant position at a higher pay grade could bring a quick raise in salary. Such positions were frequently advertised in the telegraphers' journals, and word of job openings spread quickly via the electric "grapevine" of the telegraphic network. As the telegrapher Minerva C. Smith observed in 1907, "Often the only way the operators can secure an ultimate raise without incessant and usually fruitless demands is to travel from city to city and from one company to the other."[67]

For women who considered telegraphy a career, moving around became part of a lifestyle that appears surprisingly contemporary to modern observers. A career in telegraphy took Fannie Wheeler, the station agent's daughter in Vinton, Iowa, from the local railroad depot to the California boom towns in the late 1860s and early 1870s. Her first move took her to Waterloo, Iowa; she then relocated to the Chicago Western Union office, where she managed the newly created ladies' department in 1869. Soon thereafter she headed west, first to Omaha and later San Francisco, where she worked in 1874 and 1875. In 1876 she moved down the California coast, first to Los Angeles and then to Santa Barbara. The *Vinton Eagle* for December 2, 1874, kept hometown readers apprised of her whereabouts, under the heading "What Can a Woman Do?":

> She can do very much outside of what some people choose to call a woman's sphere. We have an instance in point. . . . Miss Fannie M. Wheeler commenced operating at Blairstown, on the Northwestern road, before Vinton had a telegraph office. Subsequently she operated in other places—as Waterloo, Chicago, Omaha; being now in an office in the city of San Francisco, holding a responsible position at a good salary. Miss Fannie is regarded as one of the best operators in the country. While in Chicago Western Union (Main) office she worked the Union Stock Yard wire. And on one occasion received 140 messages, *without a single break*—a feat, which, probably, not one operator in a hundred could perform. . . . Who shall say *such accomplishments* are unladylike and unrefined.[68]

Fannie Wheeler's story illustrates the unusual freedom of action enjoyed by women operators in the nineteenth century. Their skills and the income their trade provided allowed them to change jobs and venues at will; the telegraphic network gave them information on jobs and opportunities for travel. Thus women operators were able to transcend the bounds of a patriarchal society if they chose to do so.

Her relatively high income enabled Mrs. O'Connor of the Chicago Western Union office to spend her vacation in her homeland, Ireland, in 1875. While there, she noted that the women operators in Ireland were allowed to sew or embroider at work when the lines were not busy; when she returned, she and the other women attempted to introduce the practice into the Chicago office, but management quickly forbade sewing on company time after business increased over the lines. By the 1890s, though, the practice was common enough in the New York office that it was mentioned by Martha Rayne.[69]

Another Chicago operator whose income permitted her to travel to Europe was Medora Olive Newell, a career telegrapher and manager for the Chicago wholesale district office of the Postal Telegraph Company, who, according to a 1909 *Telegraph Age* article, was "in the habit of spending her vacations abroad." In 1905, while returning from Europe, Newell happened to be on board the same ship as the Hungarian delegation to the Hague Peace Conference, whose members wished to send a wireless message to the Austro-Hungarian emperor, Franz Joseph, on his birthday. The ship's operator was unable to work the ship's wireless equipment, so Newell, who had some knowledge of wireless, volunteered to send the message. Members of the Hungarian delegation were so grateful for her services that they arranged for her to visit Hungary, where she was the guest of honor at several banquets and was allowed to preside ceremonially over a session of the parliament. The journal noted, that "Miss Newell is of course proud of the distinction that she has received, but has not had any false ambitions aroused thereby, and is well satisfied with her work as manager of one of the busiest branch offices in Chicago."[70]

RELIGIOUS, SOCIAL, AND CIVIC ORGANIZATIONS

While Ma Kiley and Ola Delight Smith had little use for organized religion, others were devout churchgoers who incorporated their religious beliefs into their everyday lives. For the Mormon operators of the Deseret Telegraph, telegraphy was a God-given calling, performed to benefit a larger community. Mrs. M. E. Randolph, a New York telegrapher, was a longtime member of Henry Ward Beecher's Plymouth Congregational Church in Brooklyn; during the Civil War, she put her abolitionist beliefs into practice by volunteering to go to Maryland and take charge of supplies for wounded Union soldiers. Mary Macaulay, like many other telegraphers of Irish descent, was a devout Catholic; she left her estate to Saint Peter's Catholic Church in Leroy, New York, upon her death in 1944.

Women telegraphers often belonged to mutual benefit organizations, which had a social focus as well as providing aid to the unemployed and ill. Mrs. M. E. Randolph belonged to Telegraphers' Aid, a group that provided help to telegraphers who were indigent or out of work.[71]

During World War I, Western Union established the Patriotic Service League on September 5, 1918, for its female employees; its initial function was to perform work in support of the war effort. Its members drilled in military-looking uniforms, held rallies and first aid classes, and knitted garments for the soldiers.

After the armistice, the Patriotic Service League continued to exist as a social and educational organization, changing its name to the Women's League of the Western Union. It sponsored amateur theatricals, dances, and lectures and held classes in knitting, home care for the sick, conversational French, dramatics, telegraphic topics, dancing, and outdoor sports. It also held telegraphic tournaments, awarding prizes for speed and accuracy. In Chicago, the Women's League held tours of nearby factories, including the Standard Oil refinery and the Wrigley chewing gum factory.[72]

Labor organizations often had a social aspect as well. While a member of the Order of Railroad Telegraphers in Atlanta, Georgia, in 1908, Ola Delight Smith organized the Dixie Twin Order Telegraphers' Club to promote social as well as political interaction between the ORT and the CTUA. She also organized the ladies' auxiliary of the ORT in 1909 and served as president for four years.[73]

Many women operators were strong supporters of women's issues such as the right to vote and workplace equality for women; it is not hard to detect an activist note in many of the letters written to the telegraph journals in support of equal rights for women operators. Sarah Bagley typified this spirit; she founded the Lowell Female Labor Reform Association in Lowell, Massachusetts, to improve working conditions in the textile mills before entering telegraphy in 1846. Mary Macaulay, a lifelong dedicated suffragist, served as secretary to Susan B. Anthony while working as a press operator for the *Rochester Post Express* in Rochester, New York.[74]

TELEGRAPHIC COMPETITIONS

Women as well as men competed in "telegraphic tournaments," contests in which prizes were awarded for sending and receiving Morse code with the greatest speed and accuracy. One such contest was the National Telegraphic

Tournament, held in New York City on the afternoon of April 10, 1890. Men and women competed separately; in the Ladies' Class, cash awards of $50, $40, and $20 were offered for the first, second, and third prizes respectively. Contestants were judged on the number of words they were able to send and receive in five minutes; eight women entered the Ladies' Class. K. B. Stephenson took the first prize by sending and receiving 217 words in five minutes, or a little over 43 words per minute; R. M. Dennis took second prize with 212 words, and E. R. Vanselow came in third with 210 words. The overall winner of the National Telegraphic Tournament was B. R. Pollack in the Mens' Class, who sent and received 260 words in five minutes.[75]

Later telegraphic tournaments featured receiving by typewriter as well. In a tournament held in Philadelphia in 1903, Rose Feldman of Newark, New Jersey, received a $50 first prize for the fastest receiving of twenty commercial messages on a typewriter. Although only twenty-one years old at the time, she already had nine years' experience, having begun to operate at the age of twelve in 1894.[76]

FAMILY AND MARRIAGE

The concept of "separate spheres"—the domestic sphere, centered on the home and inhabited by women and children, and the public sphere, the world of business and commerce inhabited by men—is frequently found in nineteenth-century discourses on "woman's proper place." For women operators, the debate about woman's proper place was more than an academic discussion; while male critics tended to frame the debate as a theoretical "either-or" proposition, most women telegraphers were faced with the everyday realities of trying to integrate their domestic sphere, where they kept house and raised children, with their public sphere, where they earned their livelihoods. As Carole Turbin noted in her essay on nineteenth-century working women in Troy, New York, for most working women, family life and workplace experiences are interconnected, and the relationship between them is dynamic. Women telegraphers tended to see these spheres as complementary, rather than oppositional; in her 1866 response to J. W. Stover's criticisms of women entering the public area of the telegraph office, Mrs. M. E. Lewis noted: "Woman's sphere, according to him [Stover], is keeping house. Now, exactly the qualities for a good housekeeper are those for a good telegrapher—patience, faithfulness, careful attention to numerous

and tiresome little details. If women are not on an average altogether superior to men in those qualities, I am in error."[77]

When James D. Reid, superintendent of the Atlantic and Ohio Telegraph Company, hired Emma Hunter as an operator in 1851, evidently at
her suggestion, he did not attempt to put her in a public office but instead
ran the telegraph wires into her sitting room, the proverbial center of the
domestic sphere: "We remember well that the wires were introduced into a
neat sitting room of a home in Westchester, Pa., where, with the instrument
on one side and a work basket on the other, our new assistant sent and received her messages, and filled up the interim in fixing her Sunday bonnet,
or embroidering articles of raiment which a gentleman editor is not expected to know or name."[78]

Reid's well-intentioned but chivalric attempt to preserve the public/do-
mestic dichotomy was ultimately unsuccessful; women workers, including
Hunter, soon entered the public sphere in such large numbers that the public ceased to regard their employment as novel. Ironically, though, Reid's
basically conservative gesture presaged a cultural paradigm shift far more
radical than he could have imagined; Emma Hunter, receiving and sending
messages in her sitting room in 1851, was arguably the world's first "elec-
tronic commuter."

Women Operators as Breadwinners

According to Western Union, working in the telegraph office was intended
to be a transitory phase in the lives of women, a brief interlude between
support by a father and support by a husband. A woman telegrapher's career
would be terminated by marriage, which "to them [women] is home and an
end of personal money-making for support." Thus women employees did
not need to be paid as much as men because in theory they had only themselves to support; as stated in the *Journal of the Telegraph*, "they accept terms
inferior to men because they need support only until the marriage state provides it." Reality, however, was often vastly different; as Alice Kessler-Harris
notes, "Employers freely (and largely falsely) expressed the belief that
women did not need the incomes of males because they could rely on families to support them." In fact, women who entered telegraphy generally
were working not only to support themselves but an extended family as
well; they often came from homes where the father was deceased, absent, or

irregularly employed. Anne Barnes Layton was probably typical when she reported that "I clothed and boarded myself and gave my mother $10.00 each month and helped clothe my youngest sister Sarah." A woman operator told a reporter for the *New York World* during the strike of 1883 that "there are a good many girls who work very hard and give all they earn from week to week for the support of families that are dependent on them, and of course they would suffer without work."[79]

Marital Status of Women Operators

A poem appeared in the March 6, 1875, issue of the *Telegrapher* which purported to be a tribute to all of the female operators employed in the Chicago Western Union office, written by one of their number. According to the descriptions in the verse and the accompanying text, only three of the fourteen women in the department were married; another operator was married in the interval between the time the poem was written and the time it was published. This ratio of married women to single women was probably typical for a large urban office; the majority of women operators were single and left the profession when they married.

Courtship in the New York Western Union office, however, was officially frowned upon by company management. Sensitive to their perceived in loco parentis role, telegraphic managers attempted to ban social contact between male and female operators when women were first admitted to the telegraph office. Their strictures, though, had little effect; as "J. C.," a medical doctor, observed in a letter to the *New York World* on July 24, 1883:

> Why, it used to be the rule to discharge any lady and gentleman found speaking to each other, because marriages occurred between operators more frequently than the company liked; both sexes were put under surveillance, even detectives were hired to shadow and follow operators to their homes.
>
> This element of danger, however, added zest to the undertaking, as any man who wanted to marry one of the young ladies had to use great circumspection, or lose his head in more senses than one.

During Lizzie Snow's tenure as manager of the city department in New York City, rules against social contact between male and female operators were strictly enforced. In a list of "Rules of the City Department," appearing in a letter published in the March 23, 1872, issue of the *Telegrapher*, Rule 7 stated, "Any operator corresponding, meeting, or calling at any gentleman's office, *shall forfeit her position immediately*." The letter writer added that "the mere recognition of a male acquaintance, even outside the office, and out of business hours, is punished with instant dismissal. . . . A corps of spies and detectives are employed to watch the ladies passing to and from their homes, and any infraction of the code is reported and rigorously punished."[80]

The situation at the New York Western Union office, however, appears to have been atypical and to have existed only during Snow's tenure in the 1860s and 1870s. There appears to have been no stigma attached to social contact in other offices, where male and female operators were in frequent social contact and could engage in courtship and eventually marriage without risking their livelihoods.

Career first-class operators, like Elizabeth Cogley and Mary Macaulay, often remained single for life. Many women, however, continued to work full- or part-time after marriage. Abbie Struble continued to operate on a part-time basis after her marriage to J. L. Vaughan in 1866; after his death in 1891, she went to Mexico to work full-time as an operator for the Mexican National Railroad.[81]

Some women turned to telegraphy after the death of a husband so as to support their family. Hettie Ogle became an operator in Bedford, Pennsylvania, in 1861, to support her two children after her husband, Charles, was killed in the siege of Richmond.

Women who were divorced or who had left abusive spouses also became telegraphers. Ma Kiley learned telegraphy in Del Rio, Texas, in 1901 to support her child after leaving her first husband, Paul Friesen, who had failed to support his family. Sadie Nichols, who played a major role in the strike of 1907, left Buffalo, New York, for a telegrapher's job in California after divorcing Ernest Nichols, a Buffalo police sergeant, in the late 1890s.[82]

Some of the Mormon operators of the Deseret Telegraph were involved in plural marriages. Mary Ann Johnson was the fourth wife of Aaron Johnson, one of the original settlers of Springville, Utah. After his death in 1877, she moved her family to Bancroft, Idaho, where many of her descendants became operators for the Oregon Short Line.

Emma Jane Allman was an operator in Provo and Farmington, Utah, in the 1870s who was coerced into a plural marriage, evidently against her wishes. She had been engaged to a non-Mormon man, but her father disapproved of the relationship and ordered her to terminate the engagement. Obedient to her father's will, she then became the second wife of Samuel S. Jones, a Provo businessman, in July 1878. She died in childbirth about a year later.[83]

Raising Children in the Telegraph Office

Taking care of children while holding down a full-time job was as much an issue for mothers of the nineteenth century as it is today. A reporter for the *New York World*, investigating the living conditions of New York City telegraph operators, gave this rare account of the life of a working mother in 1883:

> The reporter afterwards visited several homes of the operators. They are all in what are called "flat houses" and he found them to be as a rule what are known as genteel houses. In one case he discovered a lady with three children, who supports herself most respectably on a salary of $45 a month and sends her children to school and employs a nurse to take care of them when their mother is at work. She said, "I see very little of them. I kiss them in the morning and go away, and when I get back at night they are asleep. But to support them respectably I have to do without their society."[84]

Women operators in the West often used creative strategies to combine their domestic duties with their telegraphic work. Mary Ellen Love, an operator for the Deseret Telegraph, spent several years operating in the remote outposts of Fountain Green and Mona, Utah, before her marriage to Benjamin Barr Neff in 1870; the couple then moved to Neff's farm at Dry Creek, near Salt Lake City. The Deseret Telegraph proposed opening an office in Dry Creek following the discovery of gold at nearby Alta and offered Mary Ellen Neff the position of operator; she accepted but had the lines run to her home, enabling her to operate during her pregnancy in

1871. Two years later, with another baby only three months old, she was offered the position of operator for the Utah Central Railroad at Sandy, two miles from her home, at the generous salary of $75 a month. She accepted the job, taking the infant to work with her every day and caring for it while simultaneously managing two railroad lines and the line to the gold mine.[85]

Some women operators literally moved their families into the railroad depot so they could raise their children while on the job. Many depots had a second story that was intended for the residence of the station agent. Cassie Tomer arrived in California by covered wagon as a child and married G. W. Hill, the telegraph operator in Roseville, California, in 1876. When Hill died in 1884, leaving her a widow with five children at the age of thirty-one, she moved her family into the Roseville Depot and raised her children while serving as Wells Fargo agent, Southern Pacific ticket agent, and telegraph operator.[86]

Being raised in a telegraph office was an excellent introduction to the craft, and many children of telegraphers followed their parents into the profession as a result. Abbie Struble Vaughan's obituary describes her children's telegraphic upbringing:

> Mrs. Vaughan virtually raised her two boys, George L. and H. Latrobe, and two daughters, Madge and Lucie, in a telegraph office, for her husband followed this profession, and Mrs. Vaughan would serve as his assistant. Almost instinctively, the children took to telegraphy, and when the calls would come over the wire, and mother and father were out of the office, one of the boys or girls who happened to be near the key would recognize the signal and call to their parents to get back on the job.
>
> Subsequently "Mother" Vaughan taught her children the art of telegraphy, and each one pursued this work for a number of years.[87]

For Ma Kiley, a divorced mother of two, caring for her children while providing a living for them was a constant struggle. She took her four-year-old child, Carl, with her when she operated in Durango, Mexico, in 1903:

> I kept Carl at the depot with me for a long time, but the hotel manager where we boarded told me to leave Carl with them, say-

ing he could sleep in the same room with his son. One night around 1 A. M. I spied a little figure in a nightgown coming to the office. That hotel was fully ten blocks away and the wind was blowing a gale. Soon I recognized Carl and the guard went to meet him, asking, "No tiene Miedo, Hijite?" ("Aren't you afraid?") Carl answered, "On no, escura no come nada." ("Oh, no, the dark doesn't eat anything.")

Later, while working in Dallas in 1907, Ma Kiley was asked to go to Amarillo on a short-term assignment. Against her better judgment, and on the advice of another female telegrapher, she left Carl and her younger son, Alva Gedney, at the Episcopal Home for Children in Dallas; when she returned a week later, Alva Gedney was unconscious, with a high fever. He died three days later, at the age of two.[88]

The singularity of women telegraphers, as viewed by their contemporaries, lay not only in the arcane and technical nature of their work but in the newness of their class affiliation. They were separated from their nontelegraphic contemporaries not only by the knowledge of Morse code and office practices but also by their middle-class beliefs that a woman might acquire, through education and training, the skills necessary to lead an independent life and that there was nothing demeaning about doing so.

The appearance of this new middle-class ethos was contemporary with the establishment of telegraphy schools that admitted women, in large part because of the telegraph industry's desire to reduce labor costs. Although the availability of the necessary technical and business education made it possible for women to consider telegraphy as an occupation, it did not guarantee them a position in the industry; actually obtaining a position in a telegraph office required overcoming overt discrimination and the general male attitude that women did not belong in occupations that required them to appear in public, deal with men on an equal basis, and make judgments.

For some, telegraphy was a transitional life phase, a means of making a living between leaving the parental home and entering marriage; for others, it became a lifelong career. While some European administrations required women to resign their positions when they married, it was not uncommon for women in the United States to continue working as telegraphers after marriage; this arrangement sometimes led to their adopting creative strategies for combining domestic work and telegraphy, especially in the case of single mothers. All this led to a revaluation of the "separate spheres" ideology, at least as it applied to telegraphers, who routinely passed back and forth between the public and the domestic worlds; like Mrs. Lewis, they came to realize that "the qualities for a good housekeeper are those for a good telegrapher" and that in their lives and work experiences, the distinction between the spheres was largely artificial.

Women's Issues in the Telegraph Office

A N important aid to the study of women in the telegraph industry is
the presence of a detailed written record of the gender issues that
arose and the debate between men and women on a variety of
gender- and work-related issues, which appeared in the pages of the telegraphers' journals in the 1860s and 1870s. Often authored by women operators, these letters to the editors provide a rare view of the issues that
concerned and engaged women workers in the mid-nineteenth century.
While many of these issues were specific to the employment of women in
the telegraph industry, they also provide a valuable insight into the ideological debates that arose when women first began entering the workforce in
large numbers.

THE ENTRY OF WOMEN INTO TELEGRAPHY IN THE UNITED STATES AND THE DEBATE IN THE *TELEGRAPHER*

Many women became wage earners for the first time during the Civil War;
drafting men into the army created job vacancies, and the women they left
behind needed income to support themselves and their children. More than
one hundred thousand new jobs were opened up for women during the war,
particularly in mills, factories, and arsenals.[1]

During the Civil War, women were welcomed into the telegraph industry, often at wages equivalent to those of male operators, because of the

temporary shortage caused by the absence of men. As male telegraphers enlisted or were drafted into the Military Telegraph Corps during the Civil War, they were replaced by women in many offices. Elizabeth Cogley's skill and seven years' experience earned her a promotion to a position at Pennsylvania Railroad headquarters in Harrisburg in 1862, where, according to her obituary, written sixty years later, "expert and reliable operators were called to meet the important demands of the service." Abbie Struble, who studied telegraphy at a school set up by the Baltimore and Ohio Railroad in Pittsburgh in the 1860s, found her skills to be in demand as well; she was one of the earliest operators to learn to receive by sound alone. She operated for the Baltimore and Ohio during the war and, according to her obituary, "was credited during the Civil War with many acts of heroism," though no record of her wartime service survives.[2]

Many put their telegraphic skills to work for the Union cause out of patriotic motives. While working as a telegraph operator in Massachusetts in 1862, Mrs. M. E. Randolph heard messages pass over her line about the care and treatment of wounded soldiers; she volunteered to go to Camp Tyler, near Baltimore, and manage the distribution of supplies to the sick and wounded. Annette F. Telyea, a native of Kentucky, came to Readville, Massachusetts, to take charge of the telegraph office at the recruiting camp located there; she remained in charge of the station for the duration of the war.[3]

Others turned to telegraphy as a means of support after losing a husband in the war. Hettie Ogle became a professional telegrapher after the death of her husband, Charles, a former congressman who had enlisted early in the war and was killed in the siege of Richmond. She learned telegraphy at the Western Union office in Bedford, Pennsylvania, and later managed the telegraph office in Johnstown, Pennsylvania. Twenty-five years later, she would become famous for her heroism at the Johnstown Flood of 1889, in which she lost her life.[4]

In the Confederacy as well, women took charge of telegraph offices as men went off to war. Although even less is known about Confederate women operators than their northern counterparts, it appears that women worked as telegraphers and office managers in Georgia, South Carolina, Louisiana, Florida, and Alabama during the Civil War.[5]

A few women even served in the Military Telegraph Corps. Louisa Volker, a telegrapher in Mineral Point, Missouri, joined the corps in 1863 and provided intelligence to the Union forces on the movement of Confederates in the vicinity. During Sterling Price's invasion of Missouri in 1864, she remained in Mineral Point after the Union army retreated, which put her at risk of capture. She and her sister armed themselves with a shotgun and pistol and hid while Confederate troops entered the town and set fire to the railroad depot. Military telegraphers, who were civilians under military command and not part of the regular army, sought to gain recognition for their service after the war, a struggle that finally met with success in 1897, when Congress passed an act recognizing their military service. The only female military telegrapher other than Louisa Volker to receive a certificate of Honorable Service under the Congressional Act of January 26, 1897, was Mary E. Smith Buell, of Norwich, New York.[6]

As Virginia Penny had predicted, women were often able to enter the profession in a position of equality relative to men. In time, however, this led to their being viewed as competitors for jobs. As long as the national emergency justified it, the employment of women as telegraphers was seen in a positive light. When the men began to return from the war, however, the continued employment of women in the telegraph office was questioned by some men who saw them as a threat.[7]

As Philip Foner notes in his *Women and the American Labor Movement*, the debate over the presence of women in the telegraph office in the mid-1860s took place in the context of a heightened national awareness of the effects of the presence of large numbers of women in the workforce. Employers, who had initially hired women only because of the shortage of men, soon realized that they could reduce operating expenses by employing women at a lower rate of pay. Trade unions, which had previously resisted admitting women as members, came to the conclusion that the presence of nonunion women in the workforce would drive down wages for men as well unless women were admitted to the unions and demands for equal wages were made to employers. And women workers themselves began to organize to demand higher wages. One result of this newfound consciousness was the formation of the Working Women's Union in New York City in 1863, which demanded higher wages for women, particularly in the sewing trades. Another was the beginning of attempts by women workers to gain entry into the male-dominated trade unions and professional organizations.[8]

The earliest organization founded to promote the interests of the telegraphers was the National Telegraphic Union, founded in 1863. It perceived its functions as being social and economic rather than political; the preamble to its constitution exhorted telegraphers to unite "for the purpose of mutual protection in adversity, upholding and elevating the character and standing of our profession, promoting and maintaining between ourselves and our employers just, equitable, and harmonious relations, and advancing the general interests of the fraternity." The document avoided militant proclamations; the group perceived itself primarily as a professional organization, not a labor union. One of its principal tasks was the formation of a mutual benefit society to aid the indigent and unemployed.

The gender issue would be debated hotly in its publication, the *Telegrapher*, first published in September 1864. The *Telegrapher* was widely read; it started out with a circulation of twelve hundred readers, located from "Italy to the Pacific Coast, from Canada to Panama," and grew rapidly. By 1870, nearly all the eight thousand or so telegraphers in the United States read the biweekly journal, though those who worked for Western Union were careful not to read it on company time because Western Union considered the publication to be subversive.[9]

On October 31, 1864, a letter to the editor appeared in the second issue of the *Telegrapher* from "Susannah," who noted that "we—that is, your sister operators—are rapidly growing in numbers," and asked if women would be allowed to join the NTU "without marrying one of its members." The return address was that of the American Telegraph Company's operating rooms in New York, 145 Broadway, later the address of Western Union after the two companies merged in 1866.

Susannah's letter received an affirmative answer from the editor, Lewis H. Smith, who stated that "we know of no reason why she and her sisters cannot become members, providing they meet the qualifications. . . . No gentleman will dare refuse you admittance if you meet the requirements. The Union is established for Telegraphers, not persons."[10] One wonders if the letter had been planted by the editor to generate discussion or if Smith had simply encouraged the writer to send it. In any event, the letter generated a controversy that would last for many years, far past Smith's tenure as editor.

The November 28, 1864, issue of the *Telegrapher* contained a response from a male operator who signed himself "Spark." He began by stating condescendingly that he was pleased to see that the union would "welcome such of our sister telegraphers as wish to connect themselves with it," as long as they "timidly and modestly" knocked at the gate. But, he continued, "a prejudice exists in the minds of some of the male members of the profession, against the employment of ladies as operators," and he enumerated the reasons for this prejudice. He first questioned the technical skills of women operators, asserting that it is an "indisputable fact, that much the larger proportion of errors, in transmitting and receiving messages, are made by the female operators. . . . In the matter of penmanship, also, very many of them are sadly deficient." Spark's claim about the deficiencies of female operators

Figure 19. The American Telegraph Company Building, 145 Broadway, 1866. Courtesy Western Union Telegraph Company Collection, 1848–1963, Archives Center, National Museum of American History, Smithsonian Institution, SI neg. #89-12925.

reflects conditions in an era in which clerical work was still done primarily by male copyists. As Cindy Aron notes in *Ladies and Gentlemen of the Civil Service*, women were first employed as office workers by the Treasury Department during the Civil War and did not enter business offices in large numbers until the 1870s. Thus women entering telegraphy in the era before the feminization of office work were at a disadvantage because men were more likely to possess the requisite handwriting skills.

Spark also objected to "the disagreeable and affected style of transmission known as 'clipping,'" which, he claimed "appears to be a universal favorite among the sisterhood." Clipping consisted of not allowing the proper duration for dots and dashes while transmitting Morse code, thus making it difficult for the receiving operator to know whether a dot or dash had been sent.

Spark's main objection to women working as telegraphers, however, seems to have been their lack of a submissive attitude and failure to be properly deferential to their male colleagues: "I have, also, been surprised and pained to observe, in a few of them, an overbearing and uncourteous manner of transacting business over the wires, which is certainly not calculated to promote good feeling towards them among their co-laborers of the other sex." Spark concluded by saying that if women operators will correct these and numerous other shortcomings, then "there is a broad field open before them, if they will but make themselves worthy to occupy it."[11]

It was not long before women operators responded to Spark's claims. In the December 26, 1864, issue, a female operator, "134" (telegraphic slang for "Who is at the key?"), responded: "We know there is a prejudice against lady operators, and the very word he uses 'prejudice' . . . which, according to Webster means *premature opinion; injury; damage*, admits that it is merely a myth." Further, 134 noted obliquely that the acceptance of the right of women to work for a living was part of the belief system of the new middle class, to which most telegraphers aspired to belong, as opposed to the less enlightened value system of the working class, from whence most telegraphers came: "I notice that this feeling [prejudice against women operators] does not exist among those of our gentleman friends who stand highest in education and mental culture—those who oppose our interests being not of a gentler class." She pointed out that only those males who were themselves at the bottom of the skill and class hierarchy would feel threatened by the presence of women: "Those who criticise most are those who possess few superior qualities for business themselves." Identifying herself indirectly as a rural operator, 134 remarked on the visible role that women played in the operation of rural stations: "I know very little about your city operators, but I do know ladies, in our country offices, whom I would put second to none in point of business capacity, and where errors are unknown." Operating a rural telegraph office, a neglected area of women's business activity in the nineteenth century, required a wide range of business skills. A telegraph office required bookkeeping, regular remittances to corporate headquarters, filing of telegrams, and inventory of equipment. Thus, 134 contended, the predominance of women in small, rural telegraph offices, where the operator had to perform all the office functions, actually gave women a competitive edge in acquiring business skills. In closing, 134 demanded a retraction from Spark: "Are you quite sure that, taking the whole, the larger proportion of errors are made by the lady operators? . . . Sir, I insist that, in justice to all, you ought to retract from that statement."

The pseudonym she employed—"who is at the key?"—is itself a subtle comment on the gender issue. Telegraphy was inherently gender-neutral in that an operator had no way of knowing the gender of the person at the other end of the wire. Although claims were made that there was a distinctly female style of sending (and an associated higher level of errors), there are numerous anecdotal accounts of male operators who were surprised to learn that the "man" on the other end of the line was in fact a woman.

Another female operator, signing herself "Lightning," noted in the same issue of the periodical that male operators tended to be more critical of women operators than they were of each other: "We may exercise all the patience of Job twice over, and the very first time we fail to receive a message while they are thinking of sending it, what a time there is!"

The December 26 issue also contained a letter from a male operator, "T. A.," who went much further than "Spark" in opposing the entry of women into the profession of telegraphy. He noted with alarm that there were "a very large number of ladies learning the art of telegraphing . . . and are taught free of expense to them, by telegraph companies—with an ultimate view of course of their taking positions on such companies' lines, as soon as they arrive at a degree of proficiency."

T. A. referred to the American Telegraph Company's recently initiated program of training and hiring women as operators, under the leadership of its chief electrician, Marshall K. Lefferts. He predicted dire consequences for male operators as a result: "What this will lead to will be just this: ladies will work for a much lower salary than gentlemen, and will, under those circumstances, get the preference, and will gradually replace them." T. A. asserted that men should organize to protect their jobs: "What operators should do to protect themselves from 'hard times' is to keep the ladies out of the National Telegraphic Union, and also as much as possible off the lines."[12]

In the January 30, 1865, issue, "Aurora" called attention to a statement by a male operator who objected to women being admitted to the NTU "on an equal footing with present members." She noted with distaste male operators' frequent use of profanity over the wires and stated that "it *is* an objection to our being in the business, if we are supposed to be *only* on an equallity [*sic*] with many of the gentleman operators." Aurora suggested a direct, if somewhat subversive, way of dealing with offensive language over the wires: "My sisters telegraphic, let us make a rule to open the circuit when we hear anything of the kind; a few moments delay is nothing, compared to one soul's perjuring itself in this way." As another example of "coming down" to equality with men, Aurora cited "inhaling the foggy perfume of *miserable* segars," which women operators had to endure in the workplace.

While Aurora's comments reflect the "separate spheres" notion that women are morally superior to men, they also illustrate some of the cooperative strategies that women developed to deal with behavior they found objectionable. Opening the circuit to protest the use of profanity could cost the company time and money and risk losing customers' messages. Not surprisingly, Western Union quickly banned the use of profanity over the lines; Rule 34 of the 1870 edition of the Western Union rule book read, "Profane, obscene, or other ungentlemanly language will not be allowed upon the wires, nor in the offices of this Company."

Spark replied to 134 in the January 30, 1865, *Telegrapher* by first agreeing that "those who possess few superior qualifications for business" have "abundant reason to fear the competition of ladies," thus positioning himself as a member of the middle class with advanced views. While he felt that women operators were not yet technically on a par with the men, stating that "first class operators will not suffer from that cause until ladies become equally skillful," he added that "the right of the latter to compete with them will scarcely be questioned." Spark admitted that he may have made "too general an application" of his remarks on the penmanship of women operators but refused to retract his statements about clipping and courtesy over the line, citing as an example a case in which he had attempted to notify a woman operator over the line about her clipping and was told in return that "'if I couldn't read *that* writing I had better send an *operator* to the instrument.0'" If nothing else, Spark's anecdote indicates that women operators, even in the 1860s, were not particularly intimidated by criticism from their male colleagues. Spark attributed the larger number of errors allegedly made by women operators to "a constitutional tendency, in the female mind to 'jump at conclusions,'" while "a man more generally reaches his conclusions by facts or by reasoning, and is thus—telegraphically at least—less liable to err than his more impulsive sister."[13]

The February 27, 1865, issue contained a letter from "Magnetta," a woman telegrapher, who found T. A.'s statement that women should be kept out of the NTU and off the lines to be un-Christian and antiprogressive: "I asked myself, do I live in the nineteenth century . . . ? or are the days of barbarism rolling back upon us, and are we to do homage to the god of selfishness?" She asked rhetorically if T. A. had "forgotten that we live in a Christian age" and accused him of imitating the "dark days of heathenism." She questioned the assertion that "General Lefferts, in opening a way for ladies to become operators, does it from such selfish motive as those stated." She noted that supporters of women's rights, including Henry Ward Beecher and John B. Gough, "strive in their lectures to convince people of woman's proper sphere"; however, "General Lefferts does more; he gives us a helping hand, and places us where we can prove ourselves equal to the best of you, if only we persevere." Magnetta also brought up the issue of equal pay: "It is wrong, all wrong, when we fill a position as well as gentlemen could do, for us to have less paid us." Magnetta closed her letter with an angry personal attack on T. A.: "Protect from hard times—keep ladies out of Union; also off the lines! Sir! you weighed your soul in that remark! Please examine the weight closely. But how I shudder as I imagine your mother at home washing your linen, while your sister blacks your boots! . . .

If I have brought 'hard times' to your door by being allowed on the lines, I earnestly wish your soul may find ample field for expansion, and you be promoted to message boy, with a salary of fifty cents per day!"

The February 27 *Telegrapher* also carried a letter from "S. W. D.," a male telegraph manager who supported the right of women to join the NTU. He expressed sympathy for those males who saw "only future starvation for themselves" in allowing women to become telegraphers but hastened to "assure them the damage is less than they apprehend"; he added that "any attempt to keep women from earning a livelihood, as telegraph clerks, must fail." He advised male operators to remember "how numerous are the opportunities for *men* to carve their way in the world, and how few for women to render themselves independent."

S. W. D. noted that "we often hear it alleged that women lack the business capacity to successfully and satisfactorily transact the business of a Telegraph office." Showing a good grasp of practical psychology, he observed, "By many this assertion is uttered because they instinctively shrink from acknowledging any fear of the other sex; therefore, as the cry of 'incapacity,' if heeded, will secure the same result, they adopt it as the more politic."

S. W. D. explained that his opinions were based on personal experience. As superintendent of a line in the West in the late 1850s or early 1860s, he appointed five or six women as operators; "they did the business as promptly, satisfactorily, and with as little cause for complaint as did the same number of gentlemen at other points." He pointedly added, "Their offices were models of neatness;—cannot say as much for the others—they were proverbially 'on hand' during business hours, and there was never received against any one of them the first complaint for incivility to customers or wilful neglect of duty." His conclusion summarized the views of progressive men of the era with regard to women's rights:

> We confess to a leaning toward all reforms having for an object the bettering the condition of man or womankind. We believe women, equally with men, are "entitled to life, liberty, and the pursuit of happiness"—that a temporary advantage gained for ourselves, at the expense of another's rights, will ultimately prove a curse, which, as the adage has it, "will, like the chickens, come

home to roost." . . . It is rumored the Telegraphic Union is about taking official action upon this question of lady operators. We hope they will do so, and after thoroughly discussing it, place upon record their opinion as a body.

In the same issue, the periodical's editor Lewis H. Smith, evidently surprised at the conflict and divisiveness generated by the discussion of women operators, attempted to bring closure to the subject in an editorial titled "Lady Operators." Smith evidently had not anticipated the strong expressions of feeling the topic generated; he stated that he intended to "stop the discussion simply, because a further discussion will do no good to either party, but tend to excite unpleasant feelings."

Smith declared himself to be strongly in support of admitting women both to the profession of telegraphy and to the NTU; he realized, however, that "our remarks will be called radical" and added that he expected "few sympathizers of the male sex." He exhorted his fellow telegraphers to "give your sister as you would your brother a chance to earn an *honest* living," adding, with rare candor for the 1860s, that men did not "object to their [women] earning a dishonest one," that is, to their becoming prostitutes to support themselves. He noted with great foresight that "woman's rights will ere long be maintained, although now it is a by-word and reproach like all reforms, owing to the manner and material used in bringing it forth."

The *Telegrapher*'s editor commented on the limited range of employment open to women, asking rhetorically, "What, we ask, would you have woman do? They cannot all marry, nor make shirts, nor wash, nor iron, or be seamstresses." He noted the effect of the Civil War on the employment of women, stating, "This war of ours is throwing many unprotected women upon the world to gain a living by their own industry." Regarding women in the telegraph industry, Smith argued:

We know nothing of moment about telegraphing to exclude women therefrom, and we consider it the duty of every operator to give them the same chance that is accorded any one. If "lady operators" are instructed and allowed to improve there will be no danger of their depressing salaries. The great fault has been simply teaching a young lady the rudiments of the business and then

cooping her up in a room by herself or with others of her sex, away
from all chance of gaining knowledge, or emulating those who are
in the front rank. If men and women could change places, how
think you the former would come out? If we were hampered and
excluded as women have been for centuries, where would be our
boasted superiority?

Smith then discussed the issue of business skills, noting that the supposed tendency of women to "jump at conclusions" was simply a lack of fa-
miliarity with business terminology: "When a 'lady operator' translates
'feels' for the commercial term 'fob,' or 'seed' for 'C. O. D.,' simply because
she doesn't understand the meaning of either, we think it time she be permitted to take instructions from those who are experienced in 'all the arts
men practice.'"

Drawing a comparison to the working conditions of women in other
occupations, Smith stated that he had recently visited a printing business in
New York City, where he saw "a dozen or more women type-setters penned
up in a room not over ten feet square, steaming hot with feted [sic] air,"
while "in the adjoining room, (100 x 30 feet) were half a dozen men com-
positors, enjoying plenty of room and air." He added, "These women are
probably never allowed to go beyond mere composition, and are then spoken of as being unequal to their brothers, and denounced as reducing prices
of labor." Smith concluded his editorial by exhorting his readers to "compare these instances and draw your conclusions. It is easy to discover the
reason why women are said to be inferior. If you keep a human being a slave
for three score years, he may become a philosopher to a limited extent, but
will never reach perfection in the 'arts men practice.'"[14]

Lewis Smith's eloquent and impassioned defense of the right of women
to enter the occupation and join the NTU was intended in part to generate
support among the readers of the *Telegrapher* for a motion to admit women to
the NTU at its upcoming national convention. He and other NTU officials
who were sympathetic to the cause of the women telegraphers determined to
put the issue of membership for women to a vote at the NTU convention,
to be held on September 4, 1865, in Chicago. Their strategy was to push the
issue to the fore while it was still fresh in the minds of the NTU members.

At the Chicago convention, J. J. Flanagan, a delegate from the Louis-ville, Kentucky, district, introduced a motion that read, "*Resolved*, That in the opinion of the Convention, female telegraph operators, whose qualifica-tions are equal to the requirements of the Constitution, are entitled to membership in the Union." Flanagan added, "I don't think the subject needs any discussion. . . . The Constitution says, all who are sound opera-tors of printing, or paper operators of three years' standing, are entitled to membership, and don't say if it is black or white, female or male." To em-phasize the constitution's egalitarian intent, he noted that "in Savannah, Georgia, there is a black man, who is a perfect, sound operator."

James Patrick, a Philadelphia telegrapher who was the treasurer of the NTU, commented that he had never heard of the Savannah operator but knew of a "mulatto" operator in Macon, Georgia. J. W. Stover, a delegate from Boston, asked if passage of the motion would make it obligatory for the director of each "district," as the telegraphers' local organizations were called, to admit women operators. Flanagan and Patrick agreed that the reso-lution as worded did not require district directors to admit women. Patrick, however, questioned Flanagan's interpretation of the NTU constitution: "It don't say telegraph operators. It is operators, and according to that you might bring in operators of any class. It says nothing in that resolution but will allow sewing-machine operators to come in."

Stover then spoke in opposition to the motion. He stated that, in his opinion, "female operators are no honor or ornament to the profession" be-cause "their advantages in early life, and their training, is not such as to give them an acquaintance with business matters generally; the result is, con-stantly making blunders to such an extent, that I know every telegraphic su-perintendent who has had them employed on their lines in positions of importance, has found that they were not competent to perform the duties; and, as fast as possible, they are weeding them out." Patrick made a motion to table the resolution, but Flanagan asked him to withdraw his motion and put the resolution to a vote. The resolution to admit women to the NTU was defeated by a vote of twenty-one to five.[15]

In his editorial in the November 1, 1865, issue of the *Telegrapher*, Lewis Smith acknowledged that the resolution had failed but insisted that the con-stitution and bylaws did not bar admission of women. The NTU had not banned women from joining; it had simply voted against encouraging them to join. Again he urged women to test the system by applying for admission; whether any did is not recorded.

The issue was soon aired in a larger forum. The November 26, 1865, issue of the *New York Times* carried a letter from Mrs. M. E. Lewis, a New York City telegrapher, who objected strongly to Stover's remarks at the NTU convention. She posed a hypothetical question:

> Supposing your convention should do itself the honor to say it was a fit society for the admission of refined, intellectual women, and that professionally some of them were on a basis with yourselves, and should be entitled to all its advantages. Would it harm you in the least? Show me even the shadow of disadvantage it would cast upon you and I will be satisfied. On the contrary, it would raise you in the scale of being and we might benefit you as much or more than the advantages we might derive from such a member-ship. History has yet the task unperformed of recording one in-stance of woman's being a hindrance or detriment to any like organization.

Mrs. Lewis also took issue with Stover's statement that "female opera-tors are no honor or ornament to the profession": "There are lady telegra-phers who are both an honor and an ornament to the profession. Facts substantiate their success much too strongly for any false statements to in-fluence, and instead of superintendents being 'weeding them out,' they are most of them opening the door still wider for their employment, and in many cases award them the meed of being more efficient than men." Mrs. Lewis was correct in noting that the employment of women in the telegraph industry was increasing rather than declining, as Stover had alleged; the numbers of women employed as telegraphers increased steadily throughout the rest of the nineteenth century. She acknowledged that "we have not had the years of practice of some of you, and so far as the mere transmission of messages goes, average the matter." But, she continued, "for sterling principles, zeal and interest for those who employ them, steadiness during business hours, neatness in offices, polite treatment of customers, honesty in financial matters, &c., we are more than even with you."[16]

J. W. Stover defended his statements in the January 15, 1866, issue of the *Telegrapher.* He admitted that it was possible that "there can be found in the United States *two* or *three* lady telegraphers who will rank as first-class operators" but claimed that they were the exceptions that proved the rule because "as a class they [women] are not up to the requirements of good, reliable operators." To make a good telegrapher, he asserted, required "a general business knowledge, some acquaintance with electrical and natural phenomena as connected with the working of the wires, a superior capacity for reading manuscript, great patience, and close application," and, in his experience, "these qualifications are not easily found in females, and for obvious reasons." Stover's classification of these characteristics of the telegrapher's work as archetypally masculine was part of his attempt to construct telegraphy as an occupation best suited for males; with the exception of the requirement for knowledge of electricity, it was similar to the justifications used by male clerks and copyists to keep women out of their occupation.

Stover then proceeded to assert that the characteristics required of a good telegrapher fell outside of the mid-nineteenth-century definition of proper "womanly" behavior; he hinted that women who succeeded in business were probably lesbians because "a general knowledge of business matters is not a womanly possession, it does not belong to the sex, and it is only found in a few who, by some freak of nature, possess a woman's form, but the characteristics (to a great extent) of the masculine gender." He also alleged that women could succeed in the telegraph office only by sacrificing their womanly characteristics: "And just so far as a female is lacking in sensitiveness and womanly traits, is strong-minded and self-confident, will she excel as a telegrapher."

He also claimed that women's lack of business knowledge was not owing to limited educational opportunities but to poor choices in reading material: "It is not often that you find a young lady studying the foreign news, price-current, or stock-list, of a daily newspaper. If they read *newspapers* at all, their attention is generally attracted by the local items, the deaths and marriages." Stover concluded his defense by suggesting that women should confine themselves to more traditionally feminine activities: "If the young ladies of our times would strive as hard to fit themselves for good wives and good mothers, as they do to engage in masculine business occupations, it would be infinitely better for both sexes, and anti-matrimonial organizations among young men would not be necessary."[17]

Stover's response touched off another round of rebuttals from women in the *Telegrapher*. In the February 1, 1866, issue, Mrs. M. E. Lewis organized Stover's arguments into five categories and proceeded to refute them individually, noting sarcastically that she wished to "answer him as well as my inferior business understanding will permit." She emphasized two points. She first dealt with the charge that women did not understand business by suggesting that women be given more business education in the schools to enable them to lead independent lives:

> FIRST—It is men's fault if women do not understand business. If men did right, all women would be taught business enough while at school or afterwards, to fit them for managing their own affairs, if necessary, instead of the usual fate of being cheated or neglected by unfaithful trustees or dishonest lawyers and fellow-heirs and owners.

Lewis then attacked the notion that commercial activity is unwomanly by mentioning women who had succeeded in business and by citing the example of France, where women were commonly engaged in commercial activity:

> SECOND—The assertion that womanly women cannot understand business is thoroughly disproved by the fact that many of them do. Indeed, the most delicate, refined, most exquisitely feminine woman of our age—most lovely and most loved—Florence Nightingale—possesses, and has exerted prime business qualities as organizer and administrator, and that to a degree that A. T. Stewart might envy. But further, his assertion is disproved by the well-known and universal business activity of women in France. I do not mean in the field, but as sellers, managing clerks, and as book-keepers; in employments requiring exactly the qualities of a good telegraph operator.

Mrs. Lewis then used Stover's own arguments to prove that women had demonstrated business ability by running rural telegraph offices:

> THIRD—Mr. Stover admits that women may do very well in small stations. Now, a very large proportion of telegraph stations are small ones; and thus Mr. Stover, on his own showing, must con-

cede that a very large proportion of operators might properly be women.

To Stover's assertion that no telegraph superintendents or managers considered women equal, on average, to men as operators, she responded:

> FOURTH—This point is only to be decided by experience. And, as a matter of fact, the superintendents of my acquaintance are directly opposed to such ideas as Mr. Stover advances, for they find our lady operators at least as competent, faithful, and trustworthy, as gentlemen.

She replied to Stover's statement that "messages sent over the wires are of such a nature that womanly women ought not to be cognizant of them" as follows:

> FIFTH—If the business messages that pass over the wires are of a nature unfit to be known by women, this must be the fault of the operators; for no honest business is improper; but I am not sure that I know exactly what Mr. Stover means in this place.

Mrs. Lewis used a housecleaning metaphor to discuss the notion that business is too dirty for women to touch:

> SIXTH—It is unsafe to argue that business is so dirty that women must not touch it. It is like arguing that politics are so dirty that clergymen must have none. It is a mistake to drive cleanly people away from either. It would be better to help the cleanly people expel the dirt. Business settles itself. If lady telegraphers are incompetent and costly, the shrewd men who own the lines, and their executive forces, will soon find it out and dismiss them. But if they are faithful, safe, and cheap, they will employ them. In fact, the latter is the case, and the number of lady operators is really increasing.

Finally, Mrs. Lewis attacked Stover's attempt to construct telegraphy as a male occupation by pointing out that the requirements for a good telegrapher were the same as those for a good housekeeper, a proverbially female occupation:

> SEVENTH—There is another consideration, which of itself alto-gether destroys Mr. Stover's argument, such as it is. Woman's sphere, according to him, is keeping house. Now, exactly the qual-ities for a good housekeeper are those for a good telegrapher— patience, faithfulness, careful attention to numerous and tiresome little details. If women are not on an average altogether superior to men in those qualities, I am in error.[18]

Stover received a more sarcastic response in the March issue from "Josie," who hoped that "all lady telegraphers everywhere are subscribers to the *Telegrapher*. First, because of its intrinsic worth; and again, because we want them all to have the benefit of the views and assertions of Mr. J. W. Stover of Boston." She also hoped that the men were reading his views with interest, as many of them may have made the "sad mistake" of marrying a female telegrapher, "who, of necessity, lacks in sensitiveness and womanly traits, and, terrible thought, is strong-minded and self-confident."

> Nor is it at all a matter for congratulation that these, your wives, were successful operators, possessing a knowledge of business nec-essary to their profession. Oh! no, *that* proves them to be "*unwom-anly*," "possessing masculine characteristics," etc. Well, you *are* subjects for commiseration, and you have ours. Doubtless you thought you were obtaining high-principled, intelligent, loveable wives, and you *may be* yet undeceived as to that; but we pray you be so no longer, for Mr. Stover says (and he knows) that you are not to be envied.[19]

Stover's statement that telegraphic work was not suitable for "womanly women" was especially rankling to many women operators, who felt that their independence, self-confidence, and business knowledge were assets rather than detriments to their personal and family lives.

Lewis Smith resigned as editor of the *Telegrapher* in 1867; a notice in the journal stated that Smith was suffering from "softening of the brain" caused by "excessive mental labor." The *Telegrapher* continued publication as an independent periodical, and its editors, J. N. Ashley and Frank Pope, continued its sympathetic view toward women's rights. Again in 1868, they attempted to close discussion of the issue, with no more effect than Lewis Smith had seen in 1865.[20]

The NTU gradually faded out of existence in 1869, in part because of its reluctance to take a stand on political issues such as the rights of women and the gradual loss of privileges in the workplace. It was afraid of offending the powerful telegraph companies and was thus reluctant to become a militant labor organization. Eventually it lost its agenda completely, and telegraphers gravitated to organizations that were more responsive to their needs.[21]

As Melodie Andrews points out in her essay "'What the Girls Can Do': The Debate Over the Employment of Women in the Early American Telegraph Industry," the dire effects predicted by T. A. and others who warned against allowing women to become telegraphers simply did not materialize, at least not in the nineteenth century. Women entering the craft in large numbers after 1870 did not displace male operators; women supplemented the male force by primarily occupying lower-skilled positions that offered correspondingly lower pay. Yet by debating their opponents in print and insisting on their right to earn a living as telegraphers, women operators created their own legitimacy and visibility in the industry. By 1866, even before Western Union began to encourage the entry of women into the industry, female telegraphers had found a voice of their own and had established identities as skilled workers. The letters published in the *Telegrapher* provide a rare look into the gender politics of the workplace as practiced in the mid-nineteenth century.[22]

THE ENTRY OF WOMEN INTO TELEGRAPHY IN EUROPE

Women were first admitted to the telegraphic service in Europe during the 1850s and 1860s. Generally, women were allowed to become telegraphers for reasons of economy; a woman could manage a combined postal station and telegraph office, particularly in small towns and rural villages, at a lower rate of pay than a man would receive for the same work.

When the International Telegraphic Union (ITU) was formed in 1865, twelve of its twenty-two original member nations employed women as telegraphers. The experiences of these telegraph administrations with the employment of women were summarized in an 1870 issue of the *Journal International Télégraphique* and analyzed by Jeanne Bouvier in her 1930 study of women in the post, telegraph, and telephone administration, *Histoire des dames employées dans les postes, télégraphes, et téléphones de 1714 à 1929.* According to Bouvier, the French administration reported to the ITU that the unification of the postal and telegraph systems and the employment of women to manage the combined offices yielded a savings of approximately one thousand francs per station annually and provoked no dissension among the male employees: "We are happily and very agreeably surprised to confirm that the entry of women into the telegraphic service did not provoke any protest on the part of the male employees of that administration." Denmark, Switzerland, Wurtemberg, and Baden all reported results similar to those in France; Norway, which had employed women operators since 1858, reported exceptionally favorable results, noting that the examinations female applicants had to pass in French, English, and German, as well as the Scandinavian languages, made them especially well suited for duty on international lines. Only Sweden and Russia made negative comments; the Swedish administration remarked cryptically that it did not have reason to be satisfied with the manner in which the female operators discharged their tasks, while the Russian administration noted that women operators, already the equal of the men in language ability, needed only additional instruction to bring their performance in line with that of the men.[23]

One reason for the relative absence of conflict in Europe regarding the employment of women as telegraphers was that men did not see women as competitors to the extent that they did in the United States. Employment for men as well as women in the European postal service was relatively secure and less subject to layoffs and job reductions as was working in the private sector in the United States. The postal administrations viewed the admission of women to the telegraphic service as a means to expand its services at a relatively low cost; thus women operators were forced to accept wage discrimination as a condition for secure employment.

WOMEN'S ISSUES IN THE TELEGRAPH OFFICE IN THE UNITED STATES IN THE 1870S

The opening of the Cooper Institute in 1869 and the active support of Western Union improved opportunities for women to learn telegraphic skills and obtain employment in the telegraph industry in the United States in the 1870s. Although women had already succeeded in gaining a foothold in the telegraph office, the introduction of larger numbers of women into the profession and the creation of an integrated work environment generated new gender issues.

Whereas the debate in the *Telegrapher* in the 1860s had focused on the business skills required to perform telegraphic work and the ability of women to acquire them, the issues that arose in the 1870s involved the behavior of men and women working side by side in the workplace such as the working environment, vulgar language and harassment, and issues of sexuality in the workplace.

In the January 9, 1875, issue of the *Telegrapher*, under the heading "Will the Coming Operator Be a Woman?" a male operator using the pseudonym "Nihil Nameless" claimed that the work environment of telegraphers was unsuitable for women: "When Miss A—— has learned the business, she must get a situation. . . . She cannot afford to wait until one is offered, acceptable in every way. She must take such as she can get."[24]

The statement by Nihil Nameless that a beginning woman operator "cannot afford to wait" for a fully satisfactory position reflected the grim economic reality of life after the Panic of 1873, in which, according to the *Telegrapher*'s J. N. Ashley, "two applicants are found for every chance for work and wages."[25]

Nihil Nameless then enumerated the imagined horrors of a "less than satisfactory workplace environment" for a Victorian era woman:

> Suppose that happened to be at the stock yards, or at a railroad repairing shop, such as I have seen, her office will be surrounded, perhaps thronged with men of the rudest, most uncultured type, glaring on her through her window, asking her impertinent or insulting questions and giving utterance to the most shocking profanity. She must bear it; she cannot protect herself, nor punish the offenders . . . an accident occurs out on the railroad over which her line runs, she must go in the night and the storm, perhaps, and attach the instrument to the wires, and sitting there alone and unprotected, among blasphemous men, work while chilling rain drenches her, freezing as it falls. . . .

Night work must still be done then as now, and when the
women shall have absorbed *all the work* they must do *this*, too.

Although Nihil Nameless did not directly address the competition between
men and women for jobs, he brought up the issue indirectly by implying
that the telegraph office, and particularly the railroad yard, were unsuitable
environments for respectable women: "In what I have said I have not hoped
to influence any one already in the business to quit it. . . . But if I have said
what shall deter any lady who now contemplates *learning*, from entering a
profession which I consider eminently unfitted for her delicate organization
and pure character, I shall accomplish all I intended."[26]

As they had in the 1860s, women responded to these assertions in print.
In the January 23 issue, a female operator, "Aliquae," argued that the best
way for women to ensure respectful treatment in the workplace was to exhibit self-respect: "One who is naturally refined will not lose her identity by
coming in contact with others. And one who respects herself, will always be
respected." Aliquae also addressed the issue of women telegraphers dealing
directly with workmen and thereby coming in contact with the rougher elements of society: "And is it any worse to hand orders, etc. to a few workmen, than to give orders to the butcher, the baker, etc., at home? A woman
cannot stay forever penned up at home, and only go out into the world
hemmed in by a father or a brother on one side and a husband on the
other."[27]

The notion of women telegraphers "handing orders" to men was difficult for some to digest; in the course of everyday work, telegraphers had to
give orders to trainmen and line repairmen, as well as the messenger boys.
Aliquae dealt with the issue by drawing a parallel with women's domestic
duties; the railroad engineer who receives dispatch orders from a woman
should feel no more threat to his masculinity than does a butcher who receives an order from a housewife for a few cutlets.

Aliquae ignored Nihil Nameless's assertion that telegraphy is "eminently unfitted" for the "delicate organization" of women, as the increasing
numbers of women employed in the profession in the 1870s had already
disproved; she reminded her fellow correspondent that the days when
women could be "penned up at home" were over. In spite of the economic

situation, Nihil Nameless found no support from members of his own sex. "John Sterling" closed the discussion on the subject in the March 27, 1875, issue of the *Telegrapher* by stating, "Lady operators are an established fact, and whatever may be our views of the 'sphere of woman', we may as well accept the situation, and drop *that* subject."[28]

GENDERED BEHAVIOR IN THE WORKPLACE

The increasing numbers of women employed as telegraphers in the 1870s and the gradual removal of physical barriers between men and women created new gender-based workplace issues. Many of these issues centered around the rough behavior and crude language used by male operators as a means of asserting their individuality in a workplace that was becoming increasingly regimented and restrictive. According to the labor historian Stephen Meyer, it was a behavioral pattern that had its roots in the "boozing and brawling" work ethic of the Irish laborers who built the canals and railroads in the early part of the nineteenth century.[29]

When Emma Hunter was put in charge of the West Chester, Pennsylvania, telegraph office in 1851, it was hoped that the presence of women on the line would elevate the moral tone of the discourse between operators; the moral superiority of women was part and parcel of the "two spheres" ideology, and it was assumed that men would alter their behavior accordingly in the presence of women, both on the line and in the telegraph office. Women dealt with the ambiguous moral situation of the telegraph office in a variety of ways. Some, like "Aurora" and Hettie Ogle, behaved traditionally and forbade swearing in their presence; others, like Ma Kiley, learned to be "one of the boys": "I did learn to cuss. . . . All railroaders have to let off steam or blow up. We don't necessarily mean to be profane—the words just slip out."[30]

It was difficult to say when vulgarity crossed the line into harassment; this would become a major issue in the 1907 telegraphers' strike. To Western Union management, however, the rules were clear: Rule 34 of the Western Union *Rule Book* forbade the use of "Profane, obscene, or other ungentlemanly language" over the wires and in the telegraph office.[31]

The Chicago Western Union office was one of the earliest urban offices to abandon sexual segregation, and it was also the one whose work-place behavior issues were most frequently reported in the *Telegrapher*. In 1875, operator Lizzie Veazey of the Chicago office complained about the language used by telegrapher Ed Angell over the line; management listened in on the conversation, and Angell was fired as a result. Male operators complained that Angell, who otherwise had a good work record, should not have been dismissed for his first infraction of the rules; "Priscilla," another Chicago operator, supported his dismissal, stating that "any man that uses or did use such vulgar and profane language on a wire next to one worked by a young lady ought to be discharged."[32]

Another behavioral issue was drinking on the job. An operator named Hazleton of the Chicago Western Union office got into trouble in 1875 for having his lemonade "reinforced" while on a social outing with some female operators. The story, as reported in the *Telegrapher*, reflected not only the stereotype of male telegraphers as heavy drinkers but also the anxieties that many felt about exposing women to "moral corruption":

> Quite a joke is told on "Hazy." He had several ladies out for lemonade while sojourning at McGregor [telegraph office]. When he gave the boy his order for three lemonades the boy asked him, loud enough for the ladies to hear, "If he'd have a stick in one of them?" Although the ladies knew nothing about "*sticks*," still Hazy indignantly replied, "Certainly not," at the same time giving "the wink" to the boy. When the lemonades came in, the "*stick*" was so *large* that one of the ladies remarked she thought she smelt liquor, while the other one wondered why Mr. "H.'s" lemonade was so much darker than the others. This was too much for "Hazy." He immediately wanted to know of the boy why he put that liquor in his lemonade when he told him not to do so. This, of course, re-vealed the "*stick*" biz to the ladies, and when the boy replied, "*Well, you winked at me, darn you*," "the ways that are dark and the tricks that are vain" in the average young man of to-day, was very plainly manifested to the young ladies, who insist upon inquiring of "Hazy" at every opportunity "if he don't want a lemonade with a stick in it."[33]

Women viewed men's use of tobacco with distaste. "Priscilla" found the habit so disgusting in 1875 that she proposed a return to the old separation by sex: "No lady wants to go down or come up in an elevator where tobacco smoke is puffed in her face, and where the men are not particular whether they spit tobacco juice on her dress or on the floor. A lady cannot wear a dress one week in this office but what it becomes filthy with the tobacco juice which *some* of the men and boys spit on the floor. . . . Oh, for the good old days when we had an apartment to ourselves."[34]

Absenteeism

Women telegraphers were often accused of being absent from work more frequently than their male counterparts; male operators asserted that this was owing to menstrual indisposition and used it as an argument for not according women the same status as men in the operating room. According to a "veteran operator" interviewed at Long Branch, New Jersey, during the 1883 strike, "Women make telegraphy a passing matter. Matrimony is their outlook all the time. Every month they must have a rest. If they be married a confinement takes them away. Hence, there is no competition of any consequence. Women are available but not permanent."[35]

Menstrual periods may have contributed to absenteeism, but there were other reasons as well; women with children had to remain home with them when they were ill, and women had other family obligations to attend to as well. Anecdotal accounts do not indicate a vastly higher rate of absenteeism for women than for the men, stereotypically hard drinkers who frequently missed work because of hangovers. As Martha Rayne remarked in 1893, the women telegraphers' "method of spending their evenings is usually more wholesome than that of their brothers." In his 1865 letter to the *Telegrapher*, S. W. D. had attested that his female employees "were proverbially 'on hand' during business hours."[36]

In Europe, where postal administrations often kept detailed attendance records, women employees did exhibit a higher rate of absenteeism than their male counterparts. In England in 1895, male operators in the Central Telegraph Office in London averaged 8 days' absence per year, compared to 14.3 days for their female counterparts. In the Netherlands in 1897, women operators were twice as likely as men to have absences of 1 to 30 days' duration and three times as likely to have absences of 1 to 3 days.[37]

EQUAL PAY

As the economy slowed in the 1870s, Western Union began to see an advantage in hiring women. Although starting pay was not vastly different for men and women, the pay ceiling for women operators was lower, and thus female telegraphers could be retained for less money than their male counterparts. The issue of wage differentials was hotly debated in print during the second half of the nineteenth century. H. M. Cammon, who described herself as a "working woman," stated her opinion in the April 1869 issue of *Harper's:* "For if by any strange miracle, a woman chances to slip into any situation that, according to the traditions of society, should be filled by a man, we are complacently told that she is far more expert, and will do as much more work for half the wages. And why for half the wages? If a man get a thousand a year for doing certain work, why, in the name of common-sense and justice, should a woman be put off with five hundred for doing the same work faster and better? Will any body ever make that clear to a working-woman's comprehension."[38] Although Cammon was speaking about women's work in general, the argument precisely reflects the debate going on in the telegraph industry, even down to the wages and rationales. Ashley and Pope spoke to this issue in the August 3, 1872, issue of the *Telegrapher:*

> It is too true that the standard of compensation for females in the
> telegraph, as well as in other businesses, is less than for males; . . .
> it, in part, arises from the prevalent injustice, which, taking advantage of the limited range of employment open to females, and
> their necessities, forces them to accept a less compensation for
> similar services than is paid to their more fortunate brethren. . . .
>   We do not believe that ladies *desire* to labor for smaller compensation than others in the business.[39]

Thus by 1880 or so, the argument no longer had to do with whether women belonged in the telegraph industry; this question, remarked W. J. Johnston, editor of the *Operator,* in 1882, was "definitely and successfully solved." Johnston commented on the situation in Europe, where he alleged that "although a female operator receives only half as much pay as a man, she only does half as much work, and that not as thoroughly and promptly." Johnston responded that "in this country, a first-class female operator will handle satisfactorily 350 messages in a day of nine hours, and receives compensation equal to at least three-fourths of that of the male operator."[40]

The question that Johnston did not address was why the first-class women operators, presumably the equals in skill to their male counterparts, should receive only three-fourths the pay of the men. The issue became equal pay for equal work; as they acquired skill and experience, women telegraphers began to protest the wage differential. "Miss Brown," a New York telegrapher interviewed by a *New York World* reporter in 1883, was described as being "somewhat bitter" when she talked about the low wages of women operators:

> "What chance has a respectable woman to support herself at the
> wages we get? Do you wonder that girls prefer to go on the stage
> and exhibit themselves when they can get from $25 to $50 a week
> for it, and at any kind of work that requires brains and skill they
> can only make ten or twelve? Twelve dollars a week will not keep a
> woman dressed and pay her board. But you don't know how much
> pinching and twisting they do to accomplish it."[41]

Since women operators in the United States were employees of private companies, they had the right, in theory if not in practice, to negotiate with their employers for higher and more equitable wages. Occasionally this tactic was effective; in 1907, when Ma Kiley was told by her superior, the manager of the Austin, Texas, telegraph office, that he "paid the men sixty-five dollars and the women forty and fifty;" she replied that "this woman didn't work for any such salary." The manager, after seeing a sample of her expert sending ability, relented and gave her a salary of $65 a month.[42]

In other parts of the world, where women operators generally were employees of state-run postal administrations, pay inequities were often larger and more difficult to redress. Salaries for women operators in European telegraphic administrations during much of the nineteenth century tended to be about half of those paid to men, and requests for increases often required slow and cumbersome legislative action. The experience of women in the Norwegian telegraphic service was typical. During the 1890s, they petitioned the Storting, the Norwegian parliament, for wage increases on several occasions; they did not ask for equality with the men, fearing that this request would be seen as too radical. On every occasion, however, their demand was rejected. In 1898, virtually every woman in the telegraph service signed a petition to the railway committee of the Storting asking for a modest wage increase because the male telegraphers had received one the previous year. At first, the reaction was positive; a minority of railway committee members even recommended giving women operators pay equal to that of the men. The director of the telegraph service, however, instead of supporting the petition, sent a statement to parliament claiming that the women did not deserve a raise because they were not as competent as the men. After heated debate, the petition was rejected.

The women operators began to protest publicly not only the failure of parliament to grant them a raise but also the director's statement that they were less competent than the men. The female operators at the Kristiana telegraph office launched a spontaneous protest, which was reported in the newspapers. An operator who signed herself only as "female telegrapher," wrote to the telegraphers' journal, the *Telegrafbladet*, protesting that women operators were paid less than men, had virtually no opportunities for promotion, and were required to leave the service when they married; she noted that "it is perhaps doubtful who would prove the most competent, if we enjoyed the same conditions."[43]

The notion of a "proper wage" for male and female telegraphers was particularly problematical for nineteenth-century employers. Since the work performed by male and female telegraphers in each class of employment was practically identical, there was no justification, other than "custom," for paying women operators less than men. Thus persons familiar with the work, including not only the female operators themselves but also the male editors of the telegraph journals and the male leaders of the labor organizations, often called for wage equality for all telegraphers. The only argument that could be mustered in favor of wage inequality was that used by the chief of the Norwegian telegraphs: women operators deserved less pay because they did not perform the work as competently or as rapidly as men. And when a woman like Ma Kiley challenged this assertion by demonstrating a high level of competence, she would be promptly labeled the "exception that proves the rule."

WOMEN AS BUSINESS ENTREPRENEURS

A fairly high proportion of the debate in the *Telegrapher* during the 1860s and 1870s over the presence of women in the telegraph office revolved around business skills. Male operators claimed that women did not have them and were constitutionally unable to acquire them; women responded that the large number of small rural offices run by women were proof to the contrary, and, in the words of Mrs. Lewis, "If men did right, all women would be taught business enough while at school or afterwards, to fit them for managing their own affairs."[44]

Running a telegraph office was an area of business activity by women in the nineteenth century that has been largely overlooked by business historians. Although typically run as an agency rather than a proprietorship, a telegraph office required bookkeeping, regular remittances to corporate headquarters, filing of telegrams, and inventory of equipment. The office manager normally maintained a ledger of all expenditures and wages paid; each month these expenses were subtracted from the gross proceeds of the office and the difference remitted to corporate headquarters. Thus basic bookkeeping and filing skills were critical to the job; women who wished to be put in charge of an office had to acquire them either at a telegraphy school, where they were taught as part of the curriculum, or learn them on the job from an experienced office manager.

Most of the small rural offices managed by women generated little more than subsistence income for the lone operator; occasionally, however, an enterprising woman was able to grow the business into something larger. Hettie Mullen Ogle started out as the only operator in the Johnstown, Pennsylvania, Western Union office in 1869; by the time of the Johnstown Flood twenty years later, she managed a staff of three operators and an equal number of messenger boys. In addition to the telegraph office, she also managed the local telephone exchange. One of her three children, her daughter Minnie, was chief operator. Hettie Ogle was considered to be one of the best branch managers in the region; she had a reputation as a strict disciplinarian who did not allow tobacco chewing, drinking, or profanity in her office.[45]

Large commercial telegraph offices offered management opportunities as well, although Western Union's enthusiasm for promoting women to executive positions varied considerably over time. The highest position a woman could aspire to was to be appointed manager of the ladies' department or, later, the city department of a large urban telegraph office. In this position, she could expect to have as many as one hundred operators and an equal number of check-girls and check-boys reporting to her, as well as the staff of messenger boys. Lizzie Snow and Frances Dailey both served in this position at the New York Western Union office during the late nineteenth century.

Another possibility for the aspiring entrepreneur was to set up and operate a private telegraph line and compete with the industry giants by offering lower rates and faster service. Although the barriers to women entering the business world in the Gilded Age were formidable, a few women telegraphers made use of their knowledge of the telegraph business to start up independent companies. The *Telegrapher* reported on the establishment of a private line by two New York women in 1871:

> Two ladies—Miss M. L. Smith and Miss A. M. Glenson—one a native of the Sandwich Islands, the other of one of our Indian Settlements, both daughters of missionaries—towards the close of 1870 started a line of telegraph in New York from Broad Street to Union Square, with four offices in circuit—headquarters being at the Grand Central Hotel. The line was opened March first. The idea was to have a clear line to a few points, so as to send what they received promptly. These lines are succeeding in their work. They are sole proprietors, and practical operators.[46]

Women also invented and marketed telegraphic instruments. Clara M. Brinkerhoff, a former music teacher in New York City, became associated with the telegrapher George Cumming in the early 1880s. They worked on a new design for contact points for telegraph keys and were jointly issued a patent for periphery contact points in 1882; their contact point design subsequently received several design awards. The two inventors eventually founded the firm of Cumming and Brinkerhoff to market a telegraph key and other telegraphic inventions.[47]

OFFICE POLITICS: THE CASE OF LIZZIE SNOW

As Cindy Aron observes in *Ladies and Gentlemen of the Civil Service*, the entry of women into the office environment in the post–Civil War period also meant their participation in office politics. Although middle-class competitiveness was generally considered to be a male behavior in the nineteenth century, women quickly became involved in office politics and political intrigue in the telegraph office as their numbers grew and economic hard times increased competition for jobs.[48]

The focus of much of the political intrigue in the city department of the New York Western Union office at 145 Broadway in the early 1870s was Lydia H. (Lizzie) Snow, its controversial manager. Snow began her career at the American Telegraph Company before it was merged with Western Union in 1866. She rose quickly at Western Union during the late 1860s and early 1870s to the position of manager of the city department and was regarded by many in Western Union's management as a role model for women telegraphers. When the Cooper Union telegraphic school for women was opened in 1869, she was named its director.

Western Union was probably sincere in its support of Snow's career in the late 1860s and early 1870s; in its enthusiasm for cost-cutting, the company saw the employment of women as a primary goal and could hold up Lizzie Snow's career as an example of what female operators who chose not to marry and leave the profession could aspire to. James D. Reid, editor of the *Journal of the Telegraph*, went so far as to warn Western Union president William Orton to beware of a female rival for his position when he spoke on behalf of the "Ladies of the Telegraph" at the dedication of the Morse Memorial in 1871: "Are you not proud, Professor Morse, that your family are not all boys? It is not my business here to speak of woman's rights, but here is a sphere preeminently hers. Take care, President Orton—sometime you may have to sleep, like Stanton in the War Office, to keep a woman from taking your chair."[49]

Snow remained loyal to management and stayed at the key during the strike of 1870, though her allegedly dictatorial and tyrannical management of the city department was cited by women strikers as one of the reasons for the walkout. It is clear that she had developed little empathy with her female subordinates; letters critical of her management style began to appear regularly in the *Telegrapher*, and its editors, clearly on the side of her critics, began to repeat office gossip in the journal.

The February 19, 1870, issue of the *Telegrapher* related an incident in which a young woman threatened to resign rather than submit to the repressive dictates of "the mature and malevolent manageress, Miss SNOW." Snow brought in Thomas T. Eckert, general superintendent of the New York office, "to assist her in lecturing the refractory young woman." Eckert's admonitions evidently did not have the desired effect, since, according to the *Telegrapher*, "His feeling reproaches to the young women employed there at their want of consideration for the *motherly* care bestowed upon them by their *amiable* manageress are said to have affected them to tears— from suppressed laughter."[50]

A letter detailing Snow's repressive management policies appeared in the *Telegrapher* on March 23, 1872, signed by "One Who Has Been There." The writer stated that she had "been through the mill, and can speak feelingly on the subject" of the working conditions in the city department. She listed Snow's rules and regulations: reading books and newspapers during working hours was prohibited; all written messages to the operators had to be inspected by Lizzie Snow personally and delivered after 5 P.M.; no callers were allowed during working hours; all personal conversation, either in the room or over the line, was forbidden during working hours; and any operator found corresponding with, meeting, or calling at any man's office would be fired immediately. Further, the writer alleged, "A corps of spies and detectives are employed to watch the ladies passing to and from their homes, and any infraction of the code is reported and rigorously punished." Snow's relationship with Eckert was also mentioned: "The venerable metropolitan Superintendent is, if anything, in a more pitiable state of subjection than the operators. Any complaint to him of ill treatment, however outrageous, merely elicits the response, '*We must sustain our Manager.*'" The writer added cryptically, "For certain reasons he dare not do otherwise, however well disposed he might be."[51]

The notion that there was some sort of codependent relationship between Snow and Eckert seems to have been substantiated by subsequent events. In the spring of 1875, Eckert resigned his position after it became publicly known that he had conspired with Jay Gould and Thomas Edison to set up a company to compete with Western Union. Shortly thereafter, Lizzie Snow was demoted to chief operator and dismissed; according to the *Telegrapher,*

> The removal of Miss L. H. Snow from the position of chief opera-
> tor of the city wires, and manager of the female department of
> the Western Union main office in this city, which took place on
> Thursday, February 25, caused much comment among the telegra-
> phers generally. There are many versions of the matter in circula-
> tion, but it appears that the immediate cause of this action on the
> part of her superiors was her refusal to submit to and obey certain
> rules and regulations of the office, which applied to her as well as
> to the other chief operators. Miss Snow had occupied the position
> for many years, originally with the American Company, and after
> the consolidation, with the Western Union Company.
>   Miss F. L. Daly [*sic*] has been appointed to the position va-
> cated by Miss Snow.[52]

Thomas T. Eckert would return to Western Union in triumph as its president in a few years, after Jay Gould gained control of the company; Lizzie Snow, however, having fallen victim to the office politics at 145 Broadway, disappeared from public view. Nothing is known of her life after her departure from Western Union. The controversy surrounding Lizzie Snow's management style and her subsequent dismissal effectively put an end to Western Union's policy of promoting women to management posi-tions within the company. Women would continue to manage the largely female urban city departments, but a "glass ceiling" prevented them from occupying higher-level corporate positions until well into the twentieth century. Western Union's first female corporate officer was Kate O'Flanigan, who was appointed assistant secretary of the company in 1925. O'Flanigan, however, had never worked as a telegrapher; she worked her way up from a secretarial position.[53]

SEX AND MORALITY IN THE TELEGRAPH OFFICE

Although female operators were indeed an established fact by the mid-1870s, male operators were not necessarily comfortable with their presence in the telegraph office. Many men believed that allowing women to live independently and compete on an equal basis constituted a threat to the moral fiber of society. One writer complained in *Electric Age* in 1887 that allowing women equal footing in the telegraph office, "if followed to its legitimate conclusion, will break up the marriage state and result in what? community life, polygamous life, or barbarous life."[54]

In her 1993 essay on Atlanta telegrapher Ola Delight Smith, Jacqueline Dowd Hall remarks, "As featured in dime novels and union journals, they [women telegraphers] were notable less for their skills than for their unsupervised sexuality." This statement reflects a general perception that female telegraphers were somehow more knowing and worldly, particularly regarding sexual matters, than their more sheltered sisters who did not venture out into the workaday world. On one level, this belief was sublimated into the telegraphic romance view of telegraph operators as being available for romantic involvement; on another level, it sometimes resulted in accusations of promiscuity.[55]

One such case occurred in Syracuse, New York, in 1874, when Catherine Long, chief operator at the New York Central depot, returned to work after a serious illness to find that her manager, A. L. Dick, had spread a rumor that she had undergone an illegal abortion. Dick had complained of Long's alleged "immoral conduct" to his immediate superior, Superintendent Sidney E. Gifford, and had requested that Gifford fire Long to prevent her from "contaminating" the other women in the office. Gifford contacted Long's physician, Dr. R. W. Pease, who asserted that the charges were false. Long immediately filed a lawsuit against Dick, charging him with slander and asking $5,000 in damages. At the trial, Long took the witness stand to deny the allegations, and both Gifford and Dr. Pease spoke in her defense. Public sympathy was on Long's side; after only twenty minutes' deliberation, the jury awarded Long the full amount she had asked for in her suit. Dick was subsequently demoted from his managerial position.[56]

The operating environment of the railroad telegrapher contributed to the perception of the female operator as being at risk for "moral corruption." The railroad station in the late nineteenth century was located at the exact edge of respectable middle-class society; while it was a place where businessmen and even their wives and children might venture safely, it was also a place where they would encounter persons from vastly different levels of society, including common workmen, drunkards, gamblers, and prostitutes. Even today, when we speak of someone as coming from "the wrong side of the tracks," we are invoking the railroad depot symbolically as a point of social demarcation.

In many small towns, houses of prostitution were located near the railroad depot for the convenience of traveling salesmen and other itinerant men who frequented them. "Railroad hotel" became a common euphemism for such establishments. The expression "red light district," denoting a part of town where houses of prostitution existed, may have had its origin in the practice of railroad brakemen leaving their lanterns outside the door at houses of prostitution. Ma Kiley noted that a house of prostitution and a saloon were directly across the tracks from the railroad depot at Baker, Montana, when she worked there in 1908; it was, she said, the first time she had ever seen "women of that reputation."[57]

Women telegraphers working in railroad depots certainly knew that these places existed in the vicinity of railroad depots and knew what went on in them. It was the sort of knowledge of the world and its evil ways from which patriarchal society tried to shield women; female telegraphers were seen as being in danger of moral corruption simply by possessing this knowledge.

Anxieties about the corrupting influence of the working environment of female telegraphers were expressed by women reformers as well as by men. Writing to her sister Mary in 1907, Margaret Dreier Robins, president of the National Women's Trade Union League, claimed that a combination of low wages and the "easy tolerance" shown by some women operators toward the "sporting element" that frequented their offices had led them into the "red light district." She noted that in many Chicago hotels and drugstores, women were employed as telegraphers on a commission basis, with a base salary of between $5 and $20 a month, to which was added a percentage of the total business conducted. There was, however, "another side to this commission question"; the women had to endure sexual harassment from male customers in order to make a living wage. Robins explained:

> The young woman is acting as a public official, and she must, therefore, send messages even when they are accompanied by familiar attentions often forced upon her by the so called "sporting element" found at these public places. It is surely intolerable that she should find herself in her ordinary day's work in a position where she cannot rightfully resent such acts, and where her wage is dependent on the toleration she shows.
>
> We have known of some tragic instances where some of these same girls got into the "red light district" through the easy toleration they showed.

Robins added that she did not wish to discuss the issue publicly because she felt it would reflect on the character of women operators in general.[58]

Progressive era reformers like Robins often adopted the language of the antimonopolists in discussing the threat that the workaday world posed for working women; they spoke of the existence of a "vice trust," with agents lurking in the shadows of the workplace and attempting to drag young and unsuspecting women away into "white slavery." Her viewpoint was certainly influenced by her earlier work with immigrant women in New York, who were frequently lured into prostitution by men who took advantage of their poverty and lack of education. In the case of telegraphers, however, with their middle-class values and job skills, Robins seemed to suggest a more indirect road to ruin: first, seduction by members of the "sporting element" that frequented the telegraph office, and then a slow decline into the "last resort" of the fallen and disgraced woman—prostitution. Little evidence exists in the telegraphic literature to suggest that this was a common occurrence.[59]

Whether related to the work environment or not, women telegraphers sometimes exhibited an independence and assertiveness relating to control of their sexuality that put them at odds with their contemporaries. A case that attracted much public attention was that of Maggie McCutcheon, a twenty-year-old telegrapher in Brooklyn in 1886. The story appeared in

Electrical World under the heading "The Dangers of Wired Love," a reference to the popular telegraphic romance *Wired Love:*

> George W. McCutcheon, of 1204 Fulton Street, Brooklyn, has a daughter, Maggie, 20 years of age, whose ability to use a Morse telegraph instrument brought him into Justice Kenna's court, last week on a charge of threatening to blow her brains out. . . . Maggie was an expert telegrapher, made his store a local station, and had an instrument put in one corner. . . . He soon found out however, that Miss Maggie was always communicating with men at different points on the wire, and keeping up a flirtation."

Maggie became telegraphically acquainted with Frank Frisbie, a telegrapher with the Long Island Railroad. After the two telegraphers had met several times, Maggie's father became suspicious and made inquiries. He discovered that Frisbie was a married man with a family in Pennsylvania; he forbade his daughter to see Frisbie again, even sending her away to the Catskill Mountains to end the liaison. When Maggie returned, she applied to the manager of Western Union and resumed work as a telegrapher at 599 Bedford Avenue, where she could escape the watchful eye of her father. She also resumed the liaison with Frisbie, arranging to meet him at a house on Graham Avenue. "One Sunday evening, Mr. McCutcheon returned to his home to be told that Maggie had gone to the house in Graham Avenue where she formerly met Frisbie, and was not to return that evening. He started to bring her home and succeeded, but his daughter alleges that he threatened to blow her brains out, and she therefore had him arrested."[60]

Clearly Maggie was behaving in a way that went against the grain of late Victorian America, with actions spelling danger to the male-dominated world of 1886. Her story reflects the unique situation of women telegraphers in the latter part of the nineteenth century. Because of their profession, they commanded an income that permitted independent living and the ability to change employers at will; their telegraphic skills also allowed them to communicate with whomever they chose, without the chaperoning or censorship provided by parents and family. Thus they were able, in a sense, to step outside of the patriarchal society at will if they so desired. It is this ability to step in and out of the social bounds of their time that makes them seem so contemporary to us and so atypical—even dangerous—to their own contemporaries.

Although initially little notice was taken of the entry of women into the telegraph industry in the United States, a debate arose in the pages of the *Telegrapher* around the end of the Civil War over the presence of women in the telegraph office. Men expressed fears that the telegraph companies would hire women preferentially at a lower rate of pay than men and gradually drive men from the business; they also alleged that women were deficient in business skills and "constitutionally unfit" for telegraphic work. Women articulately defended their right to work in the telegraph office and questioned the notion that they were any more error-prone than their male colleagues. As it turned out, the fears of the men were largely unjustified; although large numbers of women began to enter the occupation, they generally occupied lower-status positions at lower rates of pay than the men.

As more women entered the occupation in the 1870s and segregation by sex became less common, the focus began to shift from skill-related issues to gendered workplace behavior. Women objected to drinking, smoking, and the use of vulgar language on the part of the men, for whom these behaviors were a expression of workplace autonomy in the face of increasing industrialization. As women operators acquired more skill and experience, they began to protest wage inequities. They also began to participate in office politics, which led to the downfall of Western Union manager Lizzie Snow and the imposition of a "glass ceiling" for women managers.

Women began to enter telegraphy in Europe at about the same time as in the United States. Although there was less apparent competition between male and female operators in the European posts and telegraph administrations, there was also a larger and more institutionalized pay inequity. Women who protested the pay inequity, as in Norway, often found themselves doing battle with an entrenched bureaucracy.

Toward the end of the nineteenth century, reformers began to express fears about the potential for moral corruption inherent in placing women in a railway station. While the debate generally portrayed women operators as the helpless victims of male seducers, it was in reality the ability of women operators to live independently and communicate with whomever they chose over the wire that constituted the moral threat. Their technical skills and income freed women operators from some of the constraints of a patriarchal society.

Women Telegraphers in Literature and Cinema

PORTRAYAL OF WOMEN TELEGRAPHERS IN LITERATURE

A LTHOUGH little remembered today, a considerable body of literature developed during the nineteenth century, written by both men
and women, dealing with women as telegraphers. The growing
numbers of women in telegraphy in the 1870s and the public awareness of
their role created an entirely new literary genre, the "telegraphic romance."
In either novel or short story form, this consisted of the story of a young
woman telegrapher who finds romance as well as a livelihood in her occupation. The romantic interest is often another telegrapher, and the relationship is often carried on by means of the telegraph lines. In the end, the two
lovers marry, bringing the woman's telegraphic career to an end as she becomes a devoted wife and mother.

The telegraphic romance can be seen as a form of sentimental novel of
the type that enjoyed great popularity throughout the nineteenth century as
a result of increasing literacy and a predominantly female readership. Like
other forms of sentimental literature, the stories have a strong moralistic
theme; the central characters (especially the women) must overcome adversity and demonstrate strength of character in order to qualify for the domestic blessings of home, family, and children. What makes the telegraphic
romances unconventional and "new" is the possibility of romantic involvement with an unseen stranger, carried out by means of the electric telegraph, the technological wonder of the age.[1]

The romantic possibilities of telegraphic work were noted by women telegraphers themselves. Minnie Swan, a New York telegrapher who played a leading role in the telegraphers' strike of 1883 and married another telegrapher shortly thereafter, would later remark that

> There was an atmosphere of romance surrounding the telegraph business from the time Morse perfected his method of communication. . . . It meant something, in those succeeding years, to be a telegraph operator. They were looked upon with wonder as possessing knowledge which separated them from the rest of the crowd. Passes to theatres and on all railroads, etc., were always available. This made it possible for telegraphers, with youth and the great wide world beckoning, to give ear to the siren song of adventure. Wherever one stopped he (or sometimes she) could find employment, or, barring that, friends. Many a telegraph romance, begun "over the wire," culminated in marriage.[2]

"Along the Wires"

Telegraphic romances began to appear in print in the early 1870s, coincident with the opening of the Cooper Institute and Western Union's efforts to bring more women into the industry. One of the first telegraphic romances to appear in print was Justin McCarthy's "Along the Wires," which appeared in the February 1870 issue of *Harper's Monthly Magazine*.[3] It is the story of Annette Langley, an operator in a large telegraph office in "one of the great Atlantic cities." She is an orphan; this justifies her presence in the workforce, as, through no fault of her own, "she had had to go into active busy life and work for her daily bread."

Annette occupies her time by weaving romances about the men and women who came to her office to send telegrams. One of these is a Dr. Childers, a physician who has "more theories than patients." He believes that some individuals possess a certain "sympathy" that enables them to read others deeply and see into their hearts; he suspects that this sympathy is electrical in nature. One day, Dr. Childers goes to the telegraph office to send a reply to a scientific institution in a neighboring city, where he has been asked to give a lecture. Upon handing his message to Annette, he studies

Figure 20. "Romance on the Wires," from *Harper's Weekly Magazine*, 1870s. Author's collection.

her briefly and notes that she is studying him back. He is surprised because he did not think women capable of "sympathy."

Annette, while transmitting his message—"May I venture? If so, I will come."—interprets it as having to do with a romantic liaison. He senses this and commences to send a series of mundane messages that might be interpreted as love notes. He continues to be surprised at her "sympathetic" capabilities, as he surmises that "girls of that kind are not often well trained and brought up." He is surprised that a woman of her social class possesses the ability to formulate abstractions about human nature.

Dr. Childers clearly considers himself Annette's superior, both as a male and as a member of the upper middle class, but he is not clear about her social status. As an information worker, Annette is something new, not only to Dr. Childers but also to the nineteenth century. Although she must work to support herself, her skills and literacy place her in the lower middle class, rather than the working poor.

Dr. Childers resolves to test Annette's moral character. He sends a telegram to his sister asking when her husband, of whom Dr. Childers is not overly fond, will be away, so that he can visit her. He deliberately does not mention in his telegram that the married woman to whom he is sending the message is his sister. When he hands the message blank to Annette, she blushes, proof to Dr. Childers that she interpreted his telegram as suggesting a romantic liaison with a married woman.

At this point, the author speaks directly to the reader, asking the reader to forgive Annette, a "young unmarried girl," for knowing that "there was wickedness in the world." As an orphan, "she had had to go into active busy life and work for her daily bread," unlike the reader's presumed daughters, who, being kept safely at home, could be safeguarded from any "unnecessary and premature knowledge" of the "sins and sufferings of the world." She knew about the presence of evil in the world because she "sat all day in a telegraph office, and sent along the wire any and every message which was given to her." Yet, as the author reassures the reader, "a purer heart does not beat even in your daughter's bosom than that which lived under Annette's calico gown."

Annette thinks Dr. Childers to be a cad and resolves to quit being friendly with him. She begins to distance herself from Dr. Childers, who is actually quite pleased with her rejection of his fictitious adulterous persona; it is proof of the "moral superiority" of her character, a sign of true nineteenth-century womanhood.[4]

To correct the false impression he has made, Dr. Childers then sends another telegram to his sister, this time clearly indicating in the text that it is his sister he is addressing. He is reassured by the look of surprise and relief on Annette's face; she resumes their previous informal relationship.

After a few days, Dr. Childers calls at the telegraph office and discovers that Annette has been taken ill and is in bed at her boardinghouse. He rushes to her bedside and diagnoses her as suffering from a severe "nervous attack." As she recovers, he concludes that she is in love with someone, but he cannot fathom who it could be. Annette, on her part, weeps with joy to see the doctor come to her bedside and attend her. She is genuinely touched by his concern and solicitude. Yet to her he seems preoccupied and strange; therefore, she concludes that *he* is in love with someone.

By this time, Dr. Childers has lost all sense of detachment and objectivity, yet he fails to realize that he has become obsessed with Annette. He sets out for the telegraph office one last time, determined to learn the identity of her lover. His anxiety increases as he stands in line, watching her interact with male customers. He studies her face as she talks to each and sees no reaction. Finally, it is his turn; as he approaches, she blushes. Simultaneously, they realize their attraction for each other. He tears up the message he had intended to send and hastily writes a declaration of love instead; he tells her, "I will come for an answer to-night, but not here." That evening, he proposes to her at her boardinghouse; she accepts. Soon she quits her job to become his loving wife and eventually mother to his children; she becomes a "happy, loving wife who enters with her whole soul into all the scientific theories and pursuits of her husband."

"Along the Wires" parallels nineteenth-century stories of working women in factories and mills in some respects. Annette is portrayed as a passive character who must be rescued from her situation. As Amy Gilman observes in "'Cogs to the Wheels': The Ideology of Women's Work in Mid-19th-Century Fiction," the authors of sentimental literature "turned the working woman into an idealized bourgeois heroine—a symbol of virtue and victimization."[5]

From the author's point of view, Annette's "victimization" does not consist of physical hardship like that endured by women mill workers, but rather exposure to the potential corruption incumbent upon working in the business and commercial world. Exposure to the moral vicissitudes of the commercial world had long been used as a rationale for barring women from the telegraph office.

"The Thorsdale Telegraph"

Barnet Phillips's "The Thorsdale Telegraphs," a telegraphic romance that appeared in the October 1876 *Atlantic Monthly*, deals with women operators in the railroad depot. It is the first-person story of Mary Brown, age nineteen, who has just graduated from a business college where she studied telegraphy. She reflects on the advantages of her education; were it not for her knowledge of telegraphy, she would be helping her sister teach school in a remote part of the state, something she clearly regards as a step downward.[6]

She arrives at Thorsdale, where the railroad company has offered her a position as a telegrapher. Thorsdale is a tiny midwestern town consisting of little more than the junction of two railroads. She presents herself at the telegraph office at the railroad depot. She is put off at first by Jahn Thor, the telegrapher and stationmaster, and his seeming indifference to her; he continues to stare at a timetable on the wall after she introduces herself.

After learning her name, he asks:

> "Are you a quick operator?"
> "Not very, sir; though I have had a fair practical experience."
> "So! Rather easily flustered?" he inquired, moving towards another machine, which commenced working, and over which he was now bending somewhat attentively. "Did you understand that last message?"
> "No, sir," I answered, "I was not listening; I could have understood it had I chosen to."
> "Then you are not curious?"
> "Yes, I am," I responded, by no means relishing his interrogations.

Jahn Thor is testing her. Railroad telegraphers were the air traffic controllers of the nineteenth century; they had to make split-second decisions that could affect the lives of hundreds of train passengers. As Jahn Thor would later tell Mary Brown, "You might as well at once be let into the secrets of these two railroads, both of which are shamefully mismanaged. Sometimes if it were not for promptness and decision on the part of the telegraph people, passengers would be murdered in the most wholesale way, every day in the week." Jahn Thor interrupts his interrogation of Mary Brown for a moment to listen to an incoming message. He then orders her to take the key and dictates a message for her to send: "Certain collision on or about ninety-three mile post. Keep the up-train at station, if there is time."

She leaps to the key, overturning the high stool she is sitting on in her haste, and dashes out the message. Jahn Thor is pleased; he announces triumphantly, "Correct! Miss Brown will do for an emergency." He offers her a glass of water.

She begins to suspect a hoax; she is being "hazed," a common telegraphic initiation ritual:

> "This is scandalous," I said, moving from the telegraphic table. "I do not wish your glass of water. To have imposed upon me with a message of this kind is singularly out of place. It is some stupid joke. The wires were disconnected, or led to nowhere, or you have told the receiver of the message that it amounted to nothing."

Mary Brown's outburst reveals her to be a different sort of telegrapher from the timid and passive Annette Langley. Mary Brown is a trained professional, a graduate of a telegraphic college; she considers Jahn Thor a colleague, not automatically a superior. She is comfortable with confrontation. Her attitude is a subtle comment on the change in status of women telegraphers between 1870 and 1876.

Her outburst is interrupted by an incoming message. She decodes it: "Just in time. Train stopped." The message goes on to say that an accident had been narrowly avoided. Mary Brown regrets her hasty judgment and apologizes to Jahn.

Jahn Thor, ignoring her apology, tells her irritably that the operator who sent the warning message from nearby Smoilersville is also a woman, Eusebia. He is critical of her sending practice: "Now to think of a telegraph woman sending the word 'intoxicated' when 'drunk' would have been much better, and six letters shorter. . . . She skips letters when she is the least excited. If a cow is run over, she sends me news about it, in spasms."

The reader realizes that Jahn Thor's relationship to Mary Brown, as well as with Eusebia, is somewhat competitive in nature; he is aware that women operators were taking the place of men in many railroad stations, at a lower salary.

Thor inquires after her lodging status. She has already found a good boardinghouse and has only to unpack her trunk. Thor acquaints her with his expectations and paints a grim picture of what Thorsdale has to offer in culture and entertainment, particularly for a "modern woman":

> "We might as well understand one another at once, Miss
> Brown. The office will expect your whole time and attention; and
> holidays will be few and far between. Thorsdale may have its
> faults, but there are no barbeques, nor picnics, nor bands of hope,
> nor Sisters of Washington, nor woman's rights, nor fete days, nor
> festivals here."

Jahn Thor begins to reveal something of himself. His father was a blacksmith, an immigrant from Norway, one of the original settlers of the area. His humble origins place Jahn Thor at somewhat of a social disadvantage in dealing with Mary Brown—he is in awe of her college education. But he has grandiose plans for Thorsdale. He has drawn up a master plan of the city of the future that he imagines Thorsdale will be, full of wide avenues and grand buildings.

Another message arrives from Eusebia. A drunken engineer, Kettridge, has thrown the fireman off his train and is speeding madly toward Thorsdale, out of control. If he is not stopped, a wreck is expected because there is an express train on the track on the other side of Thorsdale.

Jahn Thor asks Mary Brown to send a reply to Smoilersville: "We will do our best." When she asks what she can do to help, he crossly suggests that she shut herself in her compartment and begin plans for a "grand female university" for Thorsdale, revealing both his feeling that she was of little use in the current situation and his defensiveness about her education.

"This is trifling, sir," she retorts a bit hotly, offended but unsure of what to do next. Jahn Thor goes out to the platform, leaving her in the office alone. The train stops; Mary Brown hears a loud commotion outside. A body crashes against the office door; she opens it, afraid that something might have happened to Jahn Thor. A body falls through the door; it is the engineer Kettridge, with a pistol in his hand. As he hits the floor, the pistol discharges. If she had not opened the door precisely when she did, he likely would have shot Jahn Thor.

Thor leaps upon Kettridge, choking him. As soon as Thor has the drunken engineer subdued, Mary Brown, now fully grasping what has just transpired, falls to the floor in a faint, "in quite a resigned and satisfactory way." She is taken off to her boardinghouse in a wagon. She returns to work after a day's rest, full of doubt about her future as a telegrapher in Thorsdale; she questions whether all the stress and excitement is worth the $38 a month that the job pays.

The railroad companies, thankful for the heroic efforts of the two Thorsdale telegraphers, have arranged to send a delegation to Thorsdale to honor them with a banquet and celebration. Mary Brown finds all the publicity demeaning and wants no part of it. She insists that Thor greet the delegation, rather than herself. But she suggests to Thor that he try to persuade the railroads to aid him in his plans to develop Thorsdale. He thanks her for her advice and says, in an uncharacteristically appreciative tone, "The fact is, Miss Brown, when the general superintendent informed me that he had sent a young lady here, I was terribly opposed to it. All female telegraph operators embodied in my mind the peculiarities of the Smoilersville young lady." Mary Brown replies archly that she would enjoy having another woman operator with whom to exchange small talk and crochet patterns.

She spends the day of the celebration alone, walking along the shores of the lake, trying to sort out her feelings. She finds Thorsdale to be an impossible place, a dull rural town full of rough characters, with no particular promise. But she has to admit to a certain fondness for Jahn Thor, who for all his brusqueness was a likable person. She feels unequal to the responsibilities of the job: "A position in the office would be fraught with anxieties, and in a moment of carelessness I might be the cause of sending a whole train of cars to destruction. It was the place for a man and not for a woman."

Upon her return to the office the following day, she offers her resignation. Jahn Thor accepts her resignation, explaining that while he doubts that the events of the last few days would ever be repeated, gossips had already begun to link them romantically, and the Smoilersville paper had even reported them to be engaged.

This is too much for Mary. She seizes her hat and rushes home to her boardinghouse, where she shuts herself in her room. Some hours later, there is a knock at the door. A telegram has arrived from her sister, who is on the way to visit her. She resolves to leave Thorsdale with her sister that evening and to begin a career as a schoolteacher.

She decides to stop at the depot on the way out of town to bid farewell to Jahn Thor. He is seated alone in the telegraph office, a dim light burning. When she tells him that she is leaving and bids him good-bye, he wordlessly takes his telegraph key and taps out that she may not leave without knowing that he loves her.

Within six months, they are married. They settle in Thorsdale, which begins to grow and prosper. Soon there are children; Jahn Thor becomes a prominent citizen and is invited to attend the Centennial Exhibition in Philadelphia as a judge. Thorsdale becomes the center of prosperity and progress that Jahn Thor had dreamed of; in Mary Brown's words, "We have now rows of houses, some of them with stone fronts. We have five churches and a synagogue, six clergymen and a rabbi, ten lawyers, eleven doctors, and seventeen dentists. We have three hotels, and—to think of it!—suburban cottages. We have had a horse-race, a robbery, a divorce, a terrible fire, and municipal peculation. Are not these the attributes of a thriving town?"

The tone of "The Thorsdale Telegraphs" is light, even ironic; it is as much about the nineteenth-century faith in progress as it is about Mary Brown's romance. "The Thorsdale Telegraphs" is immediately recognizable as a more modern story than "Along the Wires"; Mary Brown, unlike Annette Langley, must choose between a career and marriage. When she accepts Jahn as her husband, she is depicted as having made the right choice; she will help him in his efforts to make Thorsdale into a modern and progressive town. Yet there is a strong subtext that says that women cannot handle the responsibilities of the telegraph office. Thus the message of the story, even with its affirmative, boosterish ending, is negative; women, even strong, assertive women like Mary Brown, do not belong in the telegraph office; they belong at home, with a family.

Josie Schofield — "Wooing by Wire"

Not surprisingly, many of the authors of telegraphic romances were women telegraphers themselves, who saw an opportunity to use their writing skills to weave a plot out of their everyday experiences at the telegraph office.

"Jo" from Toronto, Canada, already familiar to readers of the *Telegrapher* for her witty and down-to-earth letters, wrote a romance entitled "Wooing by Wire," which appeared first in the *New Dominion* of Hamilton, Ontario, and later in the *Telegrapher* in November 1875. "Jo" was actually Josie Schofield, the only woman operator at the Toronto office of the Dominion Telegraph Company in 1875.[7]

It is the story of Mildred Sunnidale, a self-described "old maid" of thirty. She has been self-supporting for twelve years, first as a schoolteacher and then as a telegrapher; her parents and other close relatives have long since died. Although "naturally of a cheerful and sociable disposition, she had always been very much alone in the world." Like Mary Brown, she prefers telegraphy to schoolteaching, which she found "wearisome enough . . . trying to train the minds of her rustic pupils." So she began to study telegraphy with the local telegraph operator after school hours. The work suited her: "With telegraphy it was different. She felt sure she could succeed at that, for her heart was in it. To her there was an interest—a fascination about it. It was so much pleasanter than going over, day after day, and year after year, the same dull lessons, with duller children. It was the height of her ambition to be put in charge of a nice little office of her own."

After applying to numerous telegraph companies and being told there were no vacancies, she confides in her telegraphic instructor that she has just about given up hope of obtaining a situation. He replies that he is sorry to hear that, as he has just decided to retire and let her take over his office. She of course takes him up on his offer; soon the wires are buzzing with the news that office "Sg" is now "manned" by a woman. Some of the operators think it quite modern and trendy to have a woman operator; others fear that the presence of a woman on the line will be a "restraint upon their freedom of (telegraphic) speech": "They would no longer be able to vent their wrath on offending brother artists in their accustomed style, which was often more forcible than elegant."

Tom Gordon, the manager of the next station on the line, was a conventional sort of fellow who, on hearing the news, was not sure "whether to be pleased at the novel idea of having a lady to work with, or to resent the innovation of a woman presuming to engage in what he considered man's work. After due consideration he came to the conclusion that it would not be likely to affect the fact to any great extent whether he resented it or not, so he philosophically resigned himself to fate, determined to make the best of it."

That evening, as Tom is having a chat with fellow operator Phil Burke, he mentions the new lady operator. Phil's first question is "Is she good looking?" "Guess so," replies the laconic Tom. "She sends well." He elaborates that so far all he has to judge her by is her sending style, which is far superior to the "nervous, jerky style" of some women. Phil says that she seems to have made quite an impression on Tom and mentions that any number of romances have begun over the telegraph wire. Tom says that that is probably as good a way as any of finding a wife: "You can form quite as good an estimate of a girl's character and temper by working and talking with her over the line as by being personally acquainted with her; better, perhaps, for you take her on her merits alone, and are not prejudiced by her appearance." After more good-natured bantering, Phil suggests that Tom may just as well propose to the young lady immediately, to avoid any prejudices acquired by actually talking to her or meeting her. Tom responds jokingly that he might just do so.

Meanwhile, Mildred Sunnidale has refurbished her office, which occupies a corner in a bookstore. She has put down a carpet and added a potted geranium and a bird cage for her canary. Telegraphers were among the earliest women to work in offices and certainly the first women workers to have their own offices. Women developed a reputation for having neater and more elaborately decorated offices than male operators; as noted in *Electrical World* in 1886: "A pretty lace or muslin curtain at the window, a bird cage hanging up aloft and some flowering plants on the narrow sill" were all sure signs of a female operator.[8]

One afternoon, as she is sitting idly, waiting for business to transact,

Tom, his curiosity aroused by his discussion with Phil, calls her office to chat a bit over the telegraph line. He asks her name and introduces himself as the operator at the next office on the line. They have no sooner gotten each other's names and signs than an "ill-natured fellow" down the line breaks in and demands that Mildred take a message. He deliberately sends it in a rapid-fire manner, making it difficult for her to keep up. She has managed to make it as far as the signature when someone slams the door of the bookstore and distracts her, causing her to miss a few characters. She "breaks" and asks him to repeat the signature, and he responds crossly that she should be reported to the superintendent and replaced with someone who won't make so many mistakes. By this time, it is fairly clear that he is trying to trip her up deliberately and "prove" that women can't handle telegraph work; it was a fairly common form of harassment of women operators.

Suddenly Tom breaks in and intervenes on her behalf. He threatens to report the man for "using insulting language and wilfully delaying business." The man relents and retransmits the signature; Mildred thinks Tom a "perfect hero" for coming to her rescue. They begin an acquaintance by telegraph, chatting over the wires during idle hours; he finds her "interesting," with "quaint little ideas" and "odd, original ways" of expressing them.

He becomes curious about her appearance; he thinks of a clever way to find out what she looks like. He will send her a photograph, and ask her to send him one of her. His plot miscarries, however, because of a misspelling. He sends the photograph, together with a note reading, "I take the liberty of sending you my photo. Please return the compliment by first mail."

Presumably he meant "complement" rather than "compliment." It was exactly the sort of mistake that male telegraphers accused women operators of making, so, although Mildred finds Tom to be handsome, she resolves to take him down a peg or two. She pretends to interpret the note literally, in which case Tom sounds a bit conceited in regarding his picture to be a "compliment." The incident is typical of "Jo's" subtle treatment of gender issues; a woman operator reading the story would understand her tactic, based on shared experience.

She sends the picture back to Tom, together with a note reading, "Many thanks for the compliment. I return it by first mail, as requested." When he receives it, he is wounded, but he blames himself for being too forward. That evening, instead of his usual good night to her before signing off, he is silent. She begins to fear that she may have been a bit too harsh with him. She closes up her office dejectedly and "walks slowly to her boarding house, feeling very wretched."

The next morning, they begin cautiously to make amends. He bids her a polite "good morning" and apologizes for being too forward in his note but feels compelled to add, "I did not think you would be so hard-hearted as to snub me so unmercifully." She replies that she only did what he asked. He answers that she knew perfectly well what he wanted her to do and offers to send his picture back to her. She does not object.

This time she keeps the picture and sends him one of herself, although not without some trepidation. Although confident that "she makes a very good picture," with her "large brown eyes" and "tip-tilted nose," she is afraid that he would be disappointed if he knew that she had red hair.

Soon after the exchange of pictures, he sends a message to Mildred, telling her that he wishes to speak with her that evening, after the offices close. He grounds the line on the other side so that only he and Mildred can communicate. He declares his love for her. She protests: "No, it is not me you love, but some ideal assemblage of virtues of your own creation, which you fancy I resemble. When you saw me, and discovered your mistake, you would repent your bargain. Why, I am an old maid, with red hair!"

He responds that red is his favorite hair color, and as for being an old maid, the older she is, the more sense she is likely to have. He arranges to come see her the following Sunday. Early Sunday morning, she rises and puts on her best muslin dress. Standing before the mirror, she is confident that she looks her best. Something begins to trouble her: it would not be honest to present herself so advantageously; she should look more ordinary, more usual. So she takes off the good dress and instead puts on a "quiet dove colored silk, that makes her look like a little Quakeress."

Soon Tom arrives, a "tall, fine looking man, with wavy brown hair, and eyes of the deepest, darkest blue." They spend the day together, walking, talking, and lunching in the parlor of the boardinghouse. By the time he leaves, they are engaged; he presents her with a ring.

Jo's story is probably at least partly autobiographical; Mildred Sunnidale presents an "internal" view of the thoughts and feelings of a woman operator, from the point of view of the operator herself. It contrasts sharply with the "external" view (how society perceived women as telegraph opera-tors) presented by the male authors of "Along the Wires" and "The Thors-dale Telegraphs."

Of the three stories, only "Wooing by Wire" discusses the issue of women in the telegraph office and how they were perceived by their male co-workers from the point of view of the operator herself. While Annette has no thoughts about her presence in a male-dominated field, and Mary Brown concludes that she has no place in the telegraph office, Mildred sim-ply gives notice that she is here and that the men should accustom them-selves to it; her self-confident attitude is owing in part to the steadily increasing numbers of women in telegraphy by the mid-1870s.

Jo's story reflects the conventional view that for most women of the age, marriage and family were the primary goals, but there is a subtext which says that having a fulfilling professional career is not incompatible with marriage and family. If one undertakes a career, it should be challenging and fulfilling in and of itself. Readers of the *Telegrapher* were already aware of Jo's views on the subject: "The grand thing for girls, as well as for men, is to find something to do, and then do it heartily. As Carlyle says, 'An endless significance lies in work: in idleness alone there is perpetual despair.' It is surely far better for us to be engaged in some useful business, earning our own living and making the most of our abilities, than to sit idly with folded hands waiting for the 'coming man', who is often so long in coming and worth so little when he arrives."[9]

All three stories suggest that women operators experienced a sense of isolation; Annette, Mary, and Mildred are all described as being "alone" to one degree or another. This derives in part, to be sure, from their socioeco-nomic status; like the majority of nineteenth-century working women, they work because they have no patriarchal family to support them and thus are lacking both familial connections and the time to enjoy a social life. Anne Barnes Layton's statement that her work caused her to "miss a lot of social life" corroborates this view. But it is also possible that their sense of isola-tion related to the technical nature of their work; few understood the work-ings of the telegraph in the mid-nineteenth century, and their technical skills must have set them apart from their contemporaries. The technical skills that separated women operators "from the rest of the crowd," to use Minnie Swan's expression, may have contributed to a sense of social alien-ation.

Ultimately, though, these stories are about marriage, "the happy ending for all good Victorian heroines," to use Amy Gilman's description. Annette achieves the most dramatic rise in social status after her marriage, as she is swept away by the presumably well-to-do Dr. Childers. Mary Brown is actually marrying down a bit when she consents to wed Jahn Thor, but she soon shares in his social ascent as Thorsdale grows and prospers. Only Mildred seems to have achieved a marriage of equality, with another operator.

Novels

Probably the most popular novel dealing with love in the telegraph office was *Wired Love,* by Ella Cheever Thayer.[10] Written in 1879, it is the story of telegrapher Nattie Rogers, her circle of friends, and her romance with the mysterious "C," an operator whom she befriends over the wires.

The story begins with Nattie, "sole presiding genius" of her telegraph office, attempting to copy incoming messages from "station Xn" while dealing with a constant barrage of questions from curious bystanders. One person interrupts to inquire if she takes the code entirely by sound; another wants to know if there is a different sound for every word or syllable. Distracted, Nattie accidentally tips over an inkwell and spills ink on her dress. Maintaining her presence of mind, she "breaks" to clean up her dress, and then sends a "G.A." (Go Ahead) to "C," the operator at Xn.

It occurs to Nattie that she is unsure of the gender of the operator at the other end—"I do wonder if this 'C' is *he* or *she*?" C, evidently wondering the same thing, asks Nattie to describe herself. She replies mischievously that she is a "tall young man"; C questions her response, claiming to be able to detect a "certain difference in the sending of a lady and a gentleman." When asked the same question by Nattie, C claims to be a "blonde, fairy-like girl," whereupon another operator breaks into the conversation to warn Nattie that C is not telling the truth about himself.

At this point, the author provides a narrative description of Nattie's background and origins. She is depicted as living in two worlds—the public world of the telegraph office, "dingy and curtailed," yet a place from whence she could "wander away, through the medium of that slender telegraph wire, on a sort of electric wings," and her private world, "bounded by the four walls of a back room at Miss Betsy Kling's," her sublet room at the Hotel Norman. We learn that she has been "compelled by the failure and subsequent death of her father to support herself"; unwilling to become a burden on her mother, Nattie "chose the more independent but harder course" of making a living as a telegrapher, for "she was not the kind of girl to sit down and wait for someone to come along and marry her, and relieve her of the burden of self-support."

Her "harder course" includes residence in a back room of a shabby boardinghouse, the Hotel Norman, where she sublets a room from Miss Betsy Kling, a gossip and would-be matchmaker who has invested considerable yet unsuccessful effort in trying to set up a romance between Nattie and another boarder, Mr. Quimby. The Hotel Norman itself teeters precariously on the edges of middle-class respectability; Nattie's window offers a view of "sheds in greater or less degree of dilapidation, a sickly grape-vine, a line of flapping sheets, an overflowing ash-barrel."

At this stage of her career, Nattie still senses "a certain fascination about telegraphy," and she clearly takes pride in being a member of an elite that understands the workings of the new technology. She is, however, well aware of the limited future that most women in the profession face, for "she had a presentiment that in time the charm would give place to monotony, more especially as, beyond a certain point, there was positively no advancement in the profession."

Returning to her residence at the Hotel Norman, Nattie is set upon by her landlady, Betsy Kling, who engages her in conversation and tries to tease out Nettie's feelings about Quimby. After escaping from Miss Kling, Nattie encounters Quimby in the hallway and flippantly asks him if the rumors she has heard about him being in love with her are true. Quimby, embarrassed, denies the stories being spread by Miss Kling. Nattie then begins to tell Quimby about C, her new acquaintance "on the wire." He is mystified by the expression, and Nattie explains the principles of telegraphy to him, noting that "no one but those who understand our language can know what we say!"

Quimby's actual love interest turns out to be Miss Cynthia Archer, another boarder at the Hotel Norman; she is an aspiring opera singer and a self-described "bohemian." Curious to meet Miss Archer, or "Cyn," as she is nicknamed, Nattie asks Quimby to bring her by the telegraph office where she works.

Shortly thereafter, Quimby and Cyn pay a visit to Nattie's office. Cyn professes ignorance about the workings of the telegraph, to which Nattie responds sharply: "Truly, the ignorance of people in regard to telegraphy is surprising; aggravating, too, sometimes." As an example of this ignorance, Nattie relates the story of a woman who came to her office to dictate a message to send. After Nattie took it down, the woman demanded to proofread it; after dotting numerous i's and crossing many t's the woman exclaimed in exasperation, "John never can read *that!* I shall have to write it myself."

After they have all had a laugh at the woman's expense, Quimby asks, "Isn't there a—a something—a *fac—similie* arrangement?" Nattie, clearly au courant with the state of the art in telegraphy, replies, "I believe there is, but it is not yet perfected." Although numerous experimental facsimile, or "fax" processes were developed during the late nineteenth century, none were commercially successful until the early twentieth century. Miss Archer then proceeds to express some trepidation at the pace of technological progress:

> "Ah, well! Then the young woman was only in advance of the age," said Miss Archer; "and what with that and the telephone, and that dreadful phonograph that bottles up all one says and disgorges at inconvenient times, we will soon be able to do everything by electricity."

Nattie's "on the wire" relationship with C continues to grow; they carry on conversations during idle moments, and Nattie finds C to be a kindred spirit in many regards; she begins to wonder what it would be like to meet him in person. Her illusions are shattered, however, when a man claiming to be C pays a visit to her office; he is overweight, with greasy hair and red-rimmed eyes that suggest a life of dissipation. He is festooned with cheap jewelry and reeks of musk. Finding him to be completely different from the mental image she had constructed, she discontinues their on-the-wire conversations and resolves to put him out of her mind.

Meanwhile, she has become fast friends with Quimby and Cyn and has become part of their "bohemian" circle of actors and musicians. They plan a "bohemian feast" together, consisting of steak, potatoes, oranges, figs, and a Charlotte russe dessert; since they have no crockery or silver, everything must be improvised, and dinner is served on boxes and chairs. Quimby brings a friend, Clem Stanwood, to the feast, who gets off to a bad start by accidentally sitting on the dessert. Clem strikes up a friendship with Nattie, who finds him to be attractive and personable; she begins to tell him of her on-the-wire relationship with C and their disastrous meeting. Clem begins to tap on the table with a pencil; Nattie soon realizes that he is sending her a message in Morse code: "Don't you 'C' the point? Can't you 'C' that you did not 'C' the 'C' you thought you did 'C' that day?" Nattie learns that Clem is the real C and that the ill-kempt and unpleasant person who had appeared at her office was an impostor.

A courtship quickly develops between Nattie and Clem, who moves in with Quimby so he can be near her. They string a telegraph line from Quimby's apartment to Nattie's so they can send love messages to each other. Betsy Kling discovers the wire, however, and demands its removal; she thinks it "immodest" for a young lady to have a telegraph wire connecting her bedroom to that of a gentleman. The story ends happily when Clem responds that Nattie is to be his wife, and they may have twelve telegraph wires set up between their bedrooms if they so desire.

Wired Love retained its popularity with telegraphers for close to two decades; it was still being advertised for sale in *Telegraph Age* in the mid-1890s. Part of the reason for the long-lived popularity of the novel was its portrayal of telegraphy as an occupation for the young and the modern. An older generation might find the unchaperoned goings-on at the Hotel Norman to be more than a bit scandalous, but younger readers could identify with the lifestyles and aspirations of the characters. *Wired Love* is one of the earliest novels to portray a unique lifestyle among young and single members of the educated middle class who have left home, have not yet married, and associate primarily with others in their own age group; in many ways, Nattie and her friends are the cultural antecedents of today's young urban professionals, or "yuppies."

The notion that telegraphers led a bohemian lifestyle did not originate with *Wired Love*; telegraphers had long had a reputation for a footloose way of life, and their high degree of literacy made them good companions for actors and artists. An "Old Telegrapher," writing in the *Telegrapher* in 1873 in a piece titled "The Peculiarities of Telegraphers in the Early and Later Periods," noted: "It must be confessed that the early telegraphers were a rather Bohemian fraternity, and they were so regarded by the community.

They were the favorites of railroad and steamboat men, and of actors and showmen, who recognized in them a congeniality of character and of pursuits which made them welcome friends and companions."[11]

While Nattie and her friends lived lives that could be described as somewhat unconventional and bohemian, Thayer was careful to show that their aspirations were totally in accord with the conventional morality of the age. While the rules might be stretched a bit by youthful exuberance, the end goal—marriage and family—was never in question.

Lida A. Churchill wrote a similar novel, *My Girls*, based on her experiences as an operator at Northbridge, Massachusetts, in 1882.[12] Churchill's main character, Cecil Emerson, is the daughter of a ne'er-do-well carpenter and Civil War veteran; her telegraphic earnings help to support her parents and her five siblings. The novel portrays the lives and experiences of a group of women telegraphers who find both successful careers and romance in New York City. Its image of the occupation and the accompanying independent lifestyle was so positive that the editors of the *Operator* felt prompted to insert the following warning in a review of the book: "We hope that no girl operators will be led by the glowing pictures that are given of the success of the heroines to think of following their example by going to New York. In this respect the story is extremely improbable, and might do harm."[13]

It is intriguing to speculate on the effect the telegraphic romances had on patterns of courtship, both among telegraphers and among the general public. Probably one direct effect was the sudden popularity of marriage by telegraph in the late 1870s. One such marriage occurred in 1876, when G. Scott Jeffreys, Western Union operator at Waynesburg, Pennsylvania, and Lida Culler, the telegrapher at Brownsville, Pennsylvania, were wed by telegraph after a courtship that began over the wires. The couple stood at the Brownsville office, together with complete wedding retinue, while the minister officiated by telegraph from the Waynesburg office. The complete text of the wedding service, including the "I do's" telegraphed by the bride and groom, was printed in the *Telegrapher*.[14]

Marriage by telegraph became such a fad that the *New York Times* felt compelled to question its legality in an editorial in 1884. While admitting that the legality of a marriage by telegraph had never been tested by the courts, the *Times* opined that "ministers who lend themselves to such a mockery of law and morality as a marriage by telegraph . . . are lending their aid to enable foolish people to live in a state of concubinage."[15]

In the Cage

Henry James's novella *In the Cage*, published in 1898, deals with issues of class and society in late Victorian England. On the surface, the story is reminiscent of "Along the Telegraph Wires." It is the story of a telegraphist in the Mayfair district of London who begins to involve her telegraphic customers in an elaborate fantasy world which she weaves out of the messages they send. She remains curiously nameless and enigmatic throughout the story—James refers to her as "our young woman."[16]

For "our young woman," the cage, the post and telegraph office she occupies in the corner of Cocker's, a grocer's store in a fashionable part of London, does not represent a chance for social advancement; rather, it is a sign of how far she has fallen socially. She must work to support her widowed mother, who has taken to drink and with whom she shares a dingy flat, which is only referred to and never seen; there are suggestions of earlier family misfortunes. Her only confidante is a Mrs. Jordan, a vicar's widow who does floral arrangements for wealthy homes; "our young woman" trades gossip about the scandals of her telegraph customers for Mrs. Jordan's descriptions of the interiors of great houses.

Although fascinated by the seemingly carefree and scandalously immoral lives of her rich and profligate customers, "our young woman" is realistic about her own prospects. She keeps company on Sundays, her day off, with a Mr. Mudge, who manages a similar office in Chalk Garden, a less fashionable neighborhood in the northwestern part of the city. She has resigned herself to marrying the practical but unimaginative Mr. Mudge, although she seems less than enthusiastic about it.

"Our young woman's" fantasies center around the person of Captain Everard, a handsome and dissolute young rake who regularly comes to her office to arrange liaisons with his lover via the telegraph; he belongs, she notes, "supremely to the class that wired everything, even their expensive feelings." Lady Bradeen, the object of Captain Everard's attentions, was "the handsomest woman she had ever seen"; she came to the telegraph office dressed in pearls and Spanish lace. The telegraphist quickly sorts out the relationship between the two by reading their messages to each other; she spends an inordinate amount of time visualizing their encounters in Regent Street and Hyde Park and Paris. She begins to feel a certain power over them; she imagines herself a co-conspirator in their affair.

Fantasy crosses over into reality when she leaves work one evening and deliberately walks past Captain Everard's residence; she is not surprised to encounter him in the street, and he greets her. They strike up a conversation and go to a nearby park to sit on a bench. She mentions that she has not yet eaten dinner; Captain Everard stops just short of asking her to dine with him, realizing that that would be socially impossible. Sensing this, she assures him that "we do feed once" during the day. She tells him that she knows of his affair and that she will help him in any way she can; then she abruptly stands and walks away. She is secretly pleased that he has lived up to her romanticized image of him by not suggesting or doing anything she would consider vulgar.

Weeks later, Captain Everard rushes into her office, demanding to see a telegram that Lady Bradeen had sent long before; he is in some mysterious trouble, and only the information in the telegram will save him. Sensing her power over him, "our young woman" pretends not to remember and asks for more information. After tantalizing Captain Everard for some time, she writes out the information he was seeking, from memory. Overjoyed at his mysterious reprieve, Captain Everard leaves the office, never to return.

"Our young woman" later learns from Mrs. Jordan that Lady Bradeen and Captain Everard are to be married, as Lord Bradeen has conveniently died, and Captain Everard, who is revealed to be penniless and somewhat of a cad, has been rescued from some unnamed calamity by a recovered message. As the telegraphist returns home from Mrs. Jordan's, she reflects that it is time to leave Mayfair and settle down in Chalk Garden with Mr. Mudge.

The story is most interesting in its portrayal of class in late Victorian England. Unlike Annette Langley, who eventually marries the object of her fantasy, "our young woman" never seriously considers a relationship with Captain Everard; an invisible class barrier separates their worlds and, paradoxically, makes it possible for her to communicate with Captain Everard on the most intimate of terms. While Annette might aspire to pass through the relatively permeable class barriers of nineteenth-century America to become Dr. Childers's wife, Henry James's telegraphist knows that for persons of her class, "it was something to fill an office under government, and she knew but too well there were places commoner still than Cocker's"; her real escape was to be found at Chalk Farm with Mr. Mudge.

WOMEN TELEGRAPHERS IN THE CINEMA

D. W. Griffith

While working for Biograph Studios, the American director D. W. Griffith was seized by the possibilities of the image of the female operator at her lonely outpost and made it the central focus of two of his earliest feature-length films, *The Lonedale Operator* and *The Girl and Her Trust*. *The Lonedale Operator* was the earlier of the two films; it was made in 1911 and starred Blanche Sweet. She is the telegraph operator at a remote western railroad station; even the name "Lonedale" suggests her isolation. When the station is attacked by bandits, she defends herself by wrapping a monkey wrench with a handkerchief, fooling the bandits into believing that it is a gun. She successfully holds them at bay until help arrives.

The Girl and Her Trust, made the following year and starring Dorothy Bernard, is essentially a remake of *The Lonedale Operator* incorporating some of the technical advances that Griffith had made in the year between the filmings of the two pictures. The plot of *The Girl and Her Trust* is more sophisticated and has better continuity; Griffith adds a chase sequence in which a moving camera mounted on a car follows a speeding train, creating an image of great speed.

Dorothy Bernard's character, Grace the telegrapher, must deal not only with bandits, who are trying to steal a company payroll, but also with several prospective suitors. The first is a pathetic and ill-dressed character played by Griffith himself in a cameo role; he attempts to ingratiate himself with the operator by offering her a soft drink. She rebuffs his overtures, however, and orders him out of the office; clearly she has learned how to deal with what Nihil Nameless referred to as "men of the rudest, most uncultured type," who tended to hang around the local depot. She also must fend off the advances of the station agent, who attempts to steal a kiss from the unsuspecting operator when she offers him a drink of her soda.

Grace's adventure begins when she receives telegraphic notice that a large company payroll is due to arrive on the next train. She is the express agent as well as the telegrapher, a not uncommon practice in the railroad depot. She is responsible for the safekeeping of the payroll until it is picked up by company officials. The station agent offers to load a pistol for her protection before he leaves; she refuses the offer, saying that "nothing ever happens" at this station. He departs with the pistol, leaving the bullets behind.

Some tramps have arrived on the train with the payroll, however, and are plotting to break into the express box and steal the money. When Grace becomes aware of their plot, she barricades herself in the telegraph office and wires to the next station for help. The thieves cut the wires, they then try to break into her office to obtain the key to the express box, but she uses scissors and a hammer to explode one of the bullets in the door lock to make the thieves believe that she has a gun.

Meanwhile, the operator at the next station has received her plea for help and dispatches a train to rescue her; he sends a train order giving the train right-of-way over all others. The thieves, unable to open the express box without the key, decide to escape with the box and break into it elsewhere. Grace leaps onto their handcar as they speed away, determined to fulfill her duty to guard the payroll. The rescue train, with the lovelorn station agent on board, bears down on the escaping bandits; Griffith films the train chase scene with a moving camera, one of the earliest uses of this technique.

Grace, grateful for her rescue, rewards the station agent with a kiss. It is the proverbial happy ending; to paraphrase Jacqueline Hall, she has proven both her competence and her femininity, asserting both her womanliness and her place as one of "the Railroad Boys."[17]

Griffith's two Biograph films are among the earliest films that can be categorized as "westerns," and it is interesting that they both feature women as central characters. Although Grace is shown as being "vulnerable," she overcomes her vulnerability through her resourcefulness and technical knowledge.

Serials: Hazards of Helen

The *Hazards of Helen* serials, filmed between 1914 and 1917, starred first Helen Holmes and later Helen Gibson as an intrepid, horseback-riding telegraph operator who regularly performed feats of derring-do to amaze theatergoers and compete with other serial heroines, such as Pearl White of the *Perils of Pauline*. Between 1914 and 1917, 119 episodes of *Hazards* were filmed.

The Leap from the Water Tower, released in 1915, starred Helen Holmes and was filmed in scenic Cajon Pass, California. Railyard scenes were filmed in the nearby San Bernardino rail yards. In this episode, a malicious brakeman, fired from his position for drinking and fighting on the job, has sabotaged the air brake lines of the train pulled by Engine number 3001. (Engine 3001 was an Atchison, Topeka and Santa Fe steam locomotive, one of the largest ever made; it was used in several of the *Hazards of Helen* serials.) After he is injured in a train accident, the brakeman is remorseful and tells his rescuer what he has done; the man races to the telegraph office and sends an urgent telegraphic message to Helen's office.

Helen, on reading the message, realizes she must warn the train crew of the danger, so she mounts her horse and rides to the nearby water tower, used to fill the boilers of the steam engines. She climbs to the top of the water tower and, when Engine 3001 races by underneath, she jumps onto the speeding locomotive. She alerts crew members of the danger, and they quickly repair the disabled air hose.

The character of the female railroad telegrapher at her lonely outpost, fending off would-be suitors while dashing off messages and saving the day for the railroad line, was almost a perfect fit for the format of the serial. The western setting made it plausible for the heroine to ride horses, deal with men as equals, and perform all the stunts and feats required for a real cliffhanger, which would have been out of character in an urban setting.

Westerns

After the advent of sound movies and the development of the western genre, several films were made that featured women telegraphers as stock characters. Portrayal of women operators was particularly common in the 1940s and 1950s, as a sense of nostalgia for the era of the railroad and the steam locomotive began to appear on the movie screen.

Western Union was directed in Hollywood in 1941 by the German Fritz Lang, best known for his 1926 work *Metropolis*. It was a semifictional account of the building of the transcontinental telegraph in 1861 and starred Randolph Scott, Robert Young, and Virginia Gilmore. Virginia Gilmore portrays Sue Creighton, a telegrapher who is the sister of Edward Creighton, supervisor of construction for the telegraph; Creighton is played by Randolph Scott. Edward Creighton was a historical figure who was largely responsible for the building of the transcontinental telegraph, but there is no evidence that he had a sister who was a telegrapher. In any event, she is present chiefly as romantic interest, being wooed simultaneously by the strong and silent Vance Shaw (Dean Jagger) and the wealthy easterner Richard Blake (Robert Young).

The movie is badly flawed by racism and sexism; the Indians are portrayed as drunken louts who must be given electric shocks to teach them to leave the wires alone. Sue Creighton watches the wagons roll west, full of linemen eager to string the transcontinental telegraph, and remarks wistfully that "things like this make some women wish that they'd been born men."

The movie partially redeems itself through its depiction of authentic telegraphic equipment of the era; numerous genuine keys, sounders, and batteries are shown in use. Western Union provided technical assistance for the movie and gets an occasional plug when outlaws are told not to "mess with Western Union."

Overland Telegraph, filmed in 1951, a Tim Holt western, stars Gail Davis in an uncharacteristically strong role as Colleen Muldoon, a telegrapher, line stringer, and manager. She is accurately portrayed as being Irish, the daughter of telegrapher Terence Muldoon. Although flawed by a ridiculous opening scene in which she is "rescued" from atop a telegraph pole by two handsome cowboys, the movie is reminiscent of the early serials in its depiction of women operators as strong, reliant, and capable of managing. Unlike Sue Creighton, Gail Davis's character is not afraid to climb poles, fix wires, manage a station, and give orders to the linemen. She becomes manager of a telegraph station after saboteurs kill her father; later, she even leads a posse in search of the desperadoes led by Brad Roberts, played by Hugh Beaumont. Throughout the story, she remains free of romantic involvement with Tim Holt and his sidekick, Chico Rafferty.

In the 1953 movie *Kansas Pacific*, directed by Ray Nazarro, telegrapher Barbara Bruce, played by Eve Miller, has to alternate between her role as railroad section chief Cal Bruce's dutiful daughter, who must prepare meals

and do the dishes, and her role as telegrapher, providing the only link between the isolated railroad camp and the rest of the world. Thus her telegraphic skills give her a special significance and make her a central character
in the plot, in which Confederate sympathizers are sabotaging the building
of a railroad in pre–Civil War Kansas.

The plot is centered around the building of the Kansas Pacific Railroad,
anachronistically situated before the Civil War; actually, the Kansas Pacific
was built in the 1870s by Jay Gould and his associates. The movie accurately
depicts prewar "bleeding Kansas" as torn by conflict between pro-Southern
forces and Union loyalists; the Confederate sympathizers, led by William
Quantrill, of Quantrill's Raiders, are trying to sabotage the building of the
Kansas Pacific to prevent the Union from having a rail link to its western
outposts. The plot may have been suggested by the real-life story of Louisa
Volker and her experiences as a military telegrapher and operator for the St.
Louis and Iron Mountain Railroad in Civil War–era Missouri.

After a construction crew is attacked near Rockwood, Kansas, the action
switches to Washington, D.C., where General Winfield Scott orders army
engineer John Nelson, played by Sterling Hayden, to go to Kansas in the
guise of a civilian construction engineer and ensure that the railroad link is
completed on time. Back in Kansas, Barbara Bruce is in the midst of clear-
ing the dinner dishes in the boxcar home she shares with her father when
she is interrupted by a message coming over the line, announcing the immi-
nent arrival of Nelson, who will oversee construction of the railroad. Cal
Bruce reacts with disbelief; he asks Barbara if she hasn't made a mistake in
reading the message. His disbelief soon turns to anger; he fears that Nelson
is being sent to replace him.

When Nelson arrives, Barbara treats him coolly at first; when he asks
what the problem is, she replies that she cannot be expected to feel kindly
toward a man who would attempt to further his career at her father's expense. Nelson assures her that that is not his intent and hints at his real purpose, which Cal Bruce has already guessed from his army-style horseback
riding.

The sabotage attempts continue, in spite of Nelson's efforts to maintain
an armed guard around the camp. Finally, after the rail route is completed
to the Colorado border, the Confederates attack trains with artillery; Nelson responds by bringing in a train loaded with Union troops and artillery,
who quickly dispatch the Confederate irregulars.

In the closing scene, Nelson bids Barbara Bruce good-bye at the railroad station; he is heading back to Washington, where he will participate in the approaching war. It is clear that a romance is budding; Nelson promises to return as soon as hostilities have ended, and Barbara promises to wait for him.

European Cinema

Some mid-twentieth-century European films used the setting of the railway station to explore social and political themes. *Closely Watched Trains*, a 1967 Czech film directed by Jiri Menzel, is a dark comedy revolving around the problematical love life of Milos Hrma, a neophyte railroad dispatcher in the town of Kostomlaty during the World War II German occupation. A subplot involves the affair between Hrma's mentor Hubicka, a cynical womanizer, and Miss Svata, the bored depot telegrapher. When Svata's mother discovers that Hubicka has not only seduced her daughter but stamped various parts of her body with the station's official seal, she marches the young truant, who actually instigated the affair, off to the town magistrate. The story of Miss Svata is emblematic of the anxieties of nineteenth century parents and reformers in the United States, who feared that the environment of the railroad station would corrupt the morals of young women who worked there. The film's depiction of a Central European railroad depot is accurate, including the recording telegraphs.

The 1990 Hungarian movie *My Twentieth Century* is full of mysterious metaphors relating to electricity and theosophy. It is the story of identical twin sisters in fin-de-siècle Central Europe, a seductress and an anarchist who lead complex and entwined lives. The film vividly depicts the impact of the electric light in the late nineteenth century, as illuminated pageants are staged at night to demonstrate the dazzling spectacle of electric lighting. A melancholy Thomas Edison appears near the end of the film to announce the inauguration of a worldwide telegraph service; women operating printing telegraphs then relay a message instantaneously around the world.

Other contemporary films use the telegraph symbolically to evoke an earlier era, when the arrival of a telegram signified important personal news; a close-up of a ticking sounder, tapping out news from afar, suggests at once both the alienation and the connectedness of the age of global communication.

The entry of women into the telegraph industry in the late nineteenth century led to the creation of a new literary genre, the telegraphic romance. A form of sentimental fiction popular during the era, the telegraphic romance exploited both the changing values of the new middle class and the romantic possibilities of the medium itself, which permitted unchaperoned exchanges with unseen persons of the opposite sex. One of the most popular telegraphic romances, Ella Cheever Thayer's *Wired Love*, described the somewhat unconventional lifestyle of the young urban professionals of the Gilded Age. Many women telegraphers made use of their literary skills to write telegraphic romances. Henry James in his novella *In the Cage* used the genre to explore issues of class in Victorian England.

Women telegraph operators were prominently featured in early silent cinema as filmmaker D. W. Griffith exploited the image of the female telegrapher at her lonely outpost; they were also ideal characters for the format of the serial. Women telegraphers were featured in several westerns, where their ability to alternate between self-reliance and domesticity made them ideal stock characters.

While the depiction of women telegraphers was often sentimentalized in both literature and cinema, nevertheless these sources provide us with unique images of telegraphers as seen by their contemporaries. They also provide valuable insights about the ability of a new communications technology to alter patterns of socialization and courtship and even to challenge conventional morality—insights that resonate as we enter the Internet age.

Women Telegraphers and the Labor Movement

ALTHOUGH most telegraphers in the United States came from a working-class background and were familiar with the role of trade unions in improving wages and working conditions, they tended to "hang back from public demonstrations" of working-class solidarity, particularly in times of economic prosperity. Instead, they gravitated toward the behavioral patterns of the traditional middle class of clerks and officials by displaying diligence at the workplace in an effort to gain the attention of management and secure raises and promotions. It was a reasonable approach, given that telegraphers were normally salaried employees and salaries were determined on an individual basis; the validity of this strategy had been demonstrated by the example of Andrew Carnegie, who had risen from messenger boy to president of United States Steel in this very manner, as he was fond of pointing out in writings and public lectures.

In hard times, however, operators lost confidence in their ability to better their station in life through individual effort, particularly when they saw the telegraph companies grow into monopolistic giants whose profits swelled, while operators were unable to maintain a living wage or steady employment. Thus the Gilded Age, a time of panics, recessions, and economic upheavals, was also a time of labor unrest among telegraphers.

The first organization founded by telegraphers was the National Telegraphic Union, established by Northern telegraphers during the Civil War. The NTU, however, did not see itself as primarily a labor organization; it was more interested in promoting professional standards and providing benefits to members. The NTU debated vigorously over whether to admit women and finally decided not to in 1865. In the late 1860s, when telegraphers were threatened with pay reductions and a more restrictive work environment, the NTU failed to take a stand on these and similar issues, and telegraphers came to regard the organization as irrelevant to their needs. The NTU finally faded away in the late 1860s as telegraphers gravitated to more militant organizations.[1]

THE TELEGRAPHERS' PROTECTIVE LEAGUE AND THE STRIKE OF 1870

In the period immediately following the Civil War, Western Union was confronted with a variety of opportunities and crises. Flushed with the success of having managed the wartime communications network for the Union while simultaneously building its Overland Telegraph route to the Pacific, Western Union began an ambitious program of expansion by absorbing its two principal rivals, the American Telegraph Company and the United States Telegraph Company, in 1866. Although this move gave Western Union near monopoly status in the United States, it also created operational and financial problems. Some of the lines of the three companies connected the same points and were therefore redundant after the merger, thereby reducing the revenue produced per mile of line. There also was a surplus of operators, as military telegraphers returned home to look for work in a market that had been expanded by the presence of women during the war. Although from a purely business standpoint this was good, because it made it possible for Western Union to hire at a lower wage and even cut wages if it so desired, it also led to excessive competition for jobs and labor unrest. Beset with these internal problems, Western Union was also faced with an economy that was essentially flat; after the booming war years, there were no prospects for economic growth in sight. The telegraph company saw its stock price plummet as investors demanded that the company take action to bolster sagging profits. Western Union's solution to all this was to begin a program of belt-tightening in the late 1860s.

To the rank-and-file operators, this move meant longer hours at reduced wages, fewer breaks, reduced benefits, and a generally restrictive op-

erating room environment. The atmosphere in the operating rooms of
Western Union after the 1866 consolidation was evocatively described by
"Minta," who wrote to the *Telegrapher* in 1868 to describe working condi-
tions for women operators:

> You forget, Mr. Editor, that we belong to the sex that are supposed
> not to need anything to eat from 7 A.M. to 6 P.M. except, perhaps,
> to nibble a cracker or a bit of fruit. When we reach home at night,
> if we are not too utterly weary to sit up . . . there are . . . inade-
> quate salaries to be eked out with the needle. . . . My thoughts
> wander sadly to the graves of departed privileges, buried, one by
> one, since the good old days when reading, writing, or studying in
> idle moments or dull hours was not a deadly sin . . . when we had
> holidays and vacations, and railroad passes and express privileges,
> and salaries graded according to proficiency, and not deducted for
> illness caused by overwork. Now we are told that the lamented
> AMERICAN Company, which took such kindly and considerate care
> of its employees, "did business in a very loose way, without order
> or system."
>
> Now really, Mr. Editor, if *order and system* has sent telegraph
> stock from 120 down to 33, and dividends from 15 per cent down
> to *nothing*, don't you think that a somewhat smaller dose of *order
> and system* might be prescribed with safety?[2]

As noted by Minta, the American Telegraph Company, under the direction of Marshall K. Lefferts, had been highly regarded by women operators;
it was one of the first to make a practice of training and hiring women, and
many women operators owed their entry into the profession to its enlightened hiring practices. In Lefferts's obituary, which appeared in the *Telegrapher* in 1876, he was eulogized as follows:

> The employment of female labor in the telegraphic service was al-
> ways a favorite scheme of General Lefferts, and one which from
> his first appointment as engineer of the American company, was
> persistently advocated and carried out by him in spite of much
> ridicule and determined opposition. This, however, in a few years

almost wholly died away, and the justice and wisdom of the step is now generally acknowledged, even by many of its most determined opponents. In the death of General Lefferts the lady operatives of the United States have to mourn the loss of a friend, to whom they are indebted more than to any other single individual for their general admission to and recognized status in the telegraphic profession.[3]

Although the overt issues discussed in Minta's letter centered around working conditions and pay, it is not hard to detect a note of dissatisfaction with the increasing "industrialization" of the craft. Resentment over the loss of prerogatives and the loss of the sense of craftsmanship and professionalism were the underlying motivations for much of the labor activism that ensued.

On October 25, 1869, operators employed by the Franklin Telegraph Company went on strike for three days to demand higher wages. Although management tried to persuade the female operators to return to work, they sided with their male counterparts and stayed off the job until management yielded and raised salaries for all operators. The success of the labor action at the Franklin Telegraph Company and the cooperative action demonstrated by the male and female operators persuaded many in the craft that telegraphers could organize and use the threat of strikes to force concessions from the telegraph companies.[4]

As a reaction to the threats of lowered wages and the loss of workplace autonomy, telegraphers founded the Telegraphers' Protective League (TPL) in 1868. The TPL took a more aggressive stand on workplace issues than the mild-mannered NTU; the main item on its agenda was maintaining Civil War–level wages in an economy in which wages in general were falling. Western Union was clearly the target of the TPL's activities; as its constitution stated, the TPL was formed "as a protection against the aggression of this powerful accumulation of capital." It would not be inaccurate to state that the TPL was first true labor organization founded by telegraphers, since the NTU refused to address workplace issues like pay and working conditions.

The TPL was a secret society; new members were required to take an oath of loyalty and secrecy, in which they pledged to "reveal neither the names of officers nor members, nor the purposes of the society." Its first

Grand Chief Operator was Ralph Pope, later an associate of Thomas Edison; other officials were W. W. Burhans, C. J. Ryan, and J. M. Peters. The leaders of the TPL correctly suspected that Western Union would try to destroy the organization if its existence became known; TPL members communicated with one another telegraphically using cipher codes. When Western Union learned that TPL members were communicating secretly over its lines, it attempted to stamp out the practice by lowering the men's wages and threatening them with discharge.[5]

Although the TPL had no specific policies relating to women, it admitted women as members, and it is clear from Minta's letter that many women operators were in sympathy with its goals. Women telegraphers supported the TPL's position on pay and participated in the strike it called in January 1870 to protest pay levels and working conditions.

It was the first major strike for the telegraph industry, then only twenty-five years old. The strike began when four men at the San Francisco Western Union office protested an attempt to cut their wages. One of the men was a member of the TPL; he requested that the TPL be allowed to arbitrate the dispute. Instead, Western Union simply fired two of the men. When operators at Sacramento heard of the firings, they demanded the right to communicate with the San Francisco operators, using the TPL cipher code; Western Union management refused. The TPL demanded that the men be reinstated; when Western Union refused, a nationwide strike was called. The strike began in San Francisco on January 3, 1870; it quickly spread to New York, Chicago, Atlanta, and Washington. At the New York Western Union office, twelve women telegraphers joined in the strike action. In Chicago, six women operators, possibly the whole ladies' department, joined the strike. The militancy of the women, remarked the *Chicago Tribune*, was a "curious feature of the present strike." The *Tribune* also noted with surprise that the women "made common cause with the men, and also suspended work. They belong to the same association, and are governed by the same rules."[6]

The concept of women going on strike was so novel that the newspapers would not print their names; however, the six nonmanagerial female employees of the Chicago Western Union office in 1869 were Emma Stanton, Addie M. Hobbs, Mary H. Kidnay, Julia J. Wirt, Josie C. Adams, and Fide M. Curtiss.

The strike, however, was unsuccessful. Western Union called in strikebreakers, mostly unemployed railroad telegraphers, to fill the positions of the striking telegraphers. On January 18, after only two weeks, the strike was called off.

The women strikers in Chicago were evidently all replaced by strike-breakers and not rehired; none of their names appear on a list of Chicago Western Union employees in 1871. The *Telegrapher* later noted that Josie C. Adams went to work as a railroad operator for the Michigan Central Railroad after the strike; she later returned to work for Western Union in Detroit, where she died of typhoid fever in 1876.[7]

To some, including the editors of the *Telegrapher*, it appeared that Western Union was being particularly vindictive toward the women who had gone on strike. Men who had gone out on strike were being allowed to go back to work, albeit at a lower wage; the women were replaced and not re-hired. Western Union seemed to feel particularly betrayed by the women strikers, who had been the recipients of education and employment opportunities not available to many women at the time. The *Telegrapher* commented editorially: "We desire to say a word for the lady operators who have participated in this movement. With one exception, elsewhere mentioned, they have remained firm and steadfast. . . . It is understood that certain Western Union officials threaten them with special proscription." The "one exception" was Mrs. M. E. Lewis, who renounced her oath of loyalty to the TPL and returned to work; she had to endure the humiliation of having her name printed in the *Telegrapher*.[8]

On January 16, two days before the end of the strike, the striking women telegraphers of New York City published a manifesto, under the signature of "The Lady Members of the Telegraphers' Protective League," in the city papers detailing their grievances; it was reprinted in the *Telegrapher* on January 22. The statement was addressed "To The Public" and began by claiming that Western Union had forbidden the rehiring of any of the women who had participated in the strike. The reason given by the telegraph company for this "especial vindictiveness" was that "these ladies were under peculiar obligation to the company for instructions in the telegraphic art, and that their attempting to assist their brother operators in a strike was an act of ingratitude and injustice towards their employers." The authors of the manifesto denied that any such obligation existed and gave statistics on their telegraphic backgrounds to back up their assertion:

In justice to ourselves we desire to state that, among the
ladies who have recently left the employ of the Western Union
Company in this city, there are five who were taught in a school
of telegraphy conducted by the American Company (not by the
Western Union), and who have worked over five years at a little
more than half the salary paid to men and boys for precisely similar service. The remaining ladies paid for their tuition themselves,
in colleges and elsewhere, in different parts of the country. Not
one of us, as far as we are aware, has received her telegraphic education at the expense of the Western Union Company.

In addition to the issue of unequal pay, the women also protested an increasingly restrictive work environment and the dictatorial management
style of the manager of the city department, Lizzie Snow:

In addition to our insufficient compensation, we have been
obliged to submit to unnecessary restrictions, indignities and in-
sults, during business hours, and to the caprice and unreasonable
exactions of the manageress, who is practically autocratic in her
government, through her influence with the officials of the Company. A simple statement of facts in regard to our treatment
would, we are confident, satisfy any reasonable person that, aside
from our sympathy with any movement of our brother operators,
we ourselves have sufficient cause to rebel, if rebellion it may be
termed.

The manifesto's writers closed with a threat to "acquaint the public with the
real condition of affairs during the last two or three years in the ladies' department of the Western Union Telegraph Office" if the telegraph company persisted in its attempts to "specially persecute us for the part we have
taken in what we believed to be a just movement."[9]

The proclamation alternated between militancy and conciliation. Although by January 16 the striking telegraphers felt that their cause was
hopeless, it is possible that the women hoped to negotiate a return to their
previous jobs at Western Union. Perhaps they hoped that Western Union
would take the opportunity to remove the cause of much of the hostility,

Lizzie Snow, the "mature and malevolent manageress," to use the *Telegrapher*'s epithet, of the city department at the New York main office. But neither the appeal to the public nor the threat to expose Western Union's management practices seemed to have had any effect on the situation of the women operators. The January 29 issue of the *Telegrapher* advertised that "some five or six young ladies, who left the Western Union office at 145 Broadway, at the time of the strike, and who are now proscribed by the *amiable* manageress of the ladies' department, are in want of situations."[10]

The failed strike spelled the end of the short-lived Telegraphers' Protective League; it had underestimated the power of the large telegraph companies and their ability to outlast the strikers. Nevertheless, the strike of 1870 demonstrated that the male and female operators could share a common agenda and work together as members of a single labor organization. After the NTU rejected membership for women in 1865, women operators had had the option of forming their own organization; women workers in the United States had, in fact, formed the Working Women's Union in 1863. In other countries, women telegraphers occasionally did form unions separate from the men's union, as happened in Norway in 1914. But this would have had the effect of splintering the telegraphers' labor organization and reducing the ability of women to negotiate for higher wages and better working conditions. Other trade unions began to admit women as well in the late 1860s and early 1870s; the Cigar Makers International Union voted to admit women as members in 1867, and the National Typographical Union, after first rejecting the membership of women in 1867, voted to amend its charter to allow membership for women in 1872.[11]

The motives of the male-dominated trade unions in admitting women as members, however, were not totally altruistic. Male trade unionists realized that unorganized women operators could provide Western Union with an abundant supply of inexpensive labor at a lower rate than the company would have to pay union men and would eventually displace them. Alice Kessler-Harris, speaking of nineteenth-century trade unions in general, observed that "women were admitted to unions after men recognized them as competitors better controlled from within than allowed to compete from without."[12]

The women may have paid a disproportionately high price for their newfound militancy; more than the men, they were penalized by Western Union for their perceived "disloyalty." Setting a pattern that would become common in later strikes, many of the women who had worked in Western Union commercial offices before the strike became railroad operators.

WOMEN AND THE LABOR MOVEMENT IN EUROPE

Women who entered the telegraphic service in Europe were less likely to join unions and play an active role in labor organizations than their counterparts in the United States, in part because of the different nature of their employment. Women operators in European administrations tended to see themselves more as civil servants or government employees than as employees of private firms with bargaining rights, as did American telegraphers. This reluctance to organize may have also reflected the greater tendency of the European women to come from middle-class backgrounds and thus to be less likely to engage in union activity. This was seen as an advantage to their hiring by some administrations. Charles Garland wrote in 1901:

> There seems little doubt that a still further consideration with many of the administrations has been the relative docility of the women. In England and France the women are practically unorganized, whilst the men have followed the general tendency to form unions for the protection of their interests and the improvement of their position. The Italian Administration describes this quality of women thus: "The women do not generally concern themselves with political questions, and are strangers to the struggles of parties and interests. This endows them with the best qualities requisite for the telegraphic service, namely, patience, discipline and application."[13]

Strikes and labor actions did occasionally occur in the European postal administrations, although less frequently than in the United States. One such action occurred in England in 1871, after discontented telegraphers in Manchester formed the Telegraphists' Association and struck for higher wages. Fearing the effects of a public disruption, the Post Office capitu-lated, raising the maximum wages of male telegraphers to £57–68 annually and the wages of their female counterparts to £42–46.[14]

The Brotherhood of Telegraphers and the Strike of 1883

The long recession that began in the United States in September 1873 took a particular toll on the telegraphers. In 1876, Western Union cut the wages of every employee, from the president on down—a massive use of the "slid-ing scale." First-class operators making around $100 a month got a 10 per-cent reduction, while those making $50 a month had their wages cut 5 percent. Those who made less than $50 a month, however, were exempted, which somewhat reduced the sex-based inequity in pay as women typically made less than $50 a month.

While Western Union cut its employees' wages in the 1870s, it contin-ued to operate at a profit and to pay handsome dividends to its stockholders. This incensed many operators, already hard-pressed in an extremely com-petitive industry; they perceived the company's actions as stealing from the poor to pay the rich. This unrest resulted in the formation of the Brother-hood of Telegraphers in 1881 and its eventual affiliation with the Knights of Labor, a national organization of trade and labor unions. The Brotherhood of Telegraphers actively promoted the rights of women telegraphers and is-sued a call for equal pay for equal work, regardless of gender.[15]

In March 1883, the Brotherhood of Telegraphers held a secret meeting in Chicago to discuss strategy in dealing with Western Union and the other telegraph companies. The leadership decided that a list of grievances would be formulated and presented to the management of Western Union; if the demands were not met, a strike would be called. A follow-up meeting was held in Philadelphia on July 14 to discuss the grievances to be presented. Chief among them was the demand for equal pay for men and women opera-tors. According to the *Chicago Tribune*, "The section of the bill of grievances in reference to equal pay for both sexes is considered of great importance. At the present time girls employed as operators, although performing generally the same class of work as the men, receive considerable less com-pensation. Their salaries do not average more than $40 a month, and, as a general thing, the girls are enthusiastic members of the Knights of Labor and one of the most potent factors of the organization in securing new members among the male operators." Apparently in an effort to defuse the issue, Western Union gave raises of $15 a month to some of the women operators in New York at about this time.[16]

On July 16, 1883, the Executive Committee of the Brotherhood of Telegraphers, under the leadership of master workman John Campbell, presented a list of demands to Thomas T. Eckert, general manager of Western Union, in New York City. (In the absence of Norvin Green, who was in Europe at the time, Eckert was acting president as well.) These included abolition of compulsory Sunday work; reduction of day shifts to eight hours a day and night shifts to seven hours; equal pay for equal work, regardless of gender; improved working conditions for linemen and railroad telegraphers; and an across-the-board 15 percent wage increase. The demands were also presented to officers of the Mutual Union Telegraph Company, the American District Telegraph Company, and the Great Northern Telegraph Company of Canada, as well as several smaller firms, but it was taken for granted that the other companies would follow Western Union's lead.

Eckert rejected the demands, claiming that the Brotherhood did not represent a majority of Western Union employees. The Brotherhood determined to call a strike if the demands were not met within forty-eight hours.[17]

The *Chicago Tribune* interviewed telegraphers and managers in Chicago to get a feel for whether a strike would be called. A Mr. McCullough, manager of the Baltimore and Ohio telegraph office in Chicago, was confident that his twenty-two railroad operators would not strike. "They have no grievances," he opined, "as far as I know." His first-class operators all got $80 or $85 a month, he stated, except for the lone woman first-class operator, who got only $70 a month, "20 a month more than any Western Union lady operator," he hastily added, remembering that one of the strike issues was equal pay for equal work.

Upon inquiring into the situation in Milwaukee, a reporter found that the operators there had no objection to women working as operators; however, the men were concerned that a woman was employed as a night operator at nearby Brookfield Junction, working ten hours each night alone at "a lonely country station unsuitable for a lady."[18]

Western Union, evidently having failed in its attempts to win over the women operators with selective raises, began to sound a more ominous note. On July 18, the following statement appeared in the *Chicago Tribune* under the signature of Thomas T. Eckert: "The statement that both sexes shall be paid the same for like service looks to the driving of women-labor from the ranks, as, were the company to concede it, it would be to its interests to prefer men operators, who can be availed of for a greater variety of service than women operators, who may be equally capable for general duty."[19]

The Brotherhood, having heard nothing from Western Union for two days, gave the company another twenty-four hours to respond. Nothing was heard from Eckert. Finally, at 12:11 P.M. on July 19, telegrapher Frank R. Phillips of the New York Western Union office climbed atop a table and blew several blasts on a whistle, signaling the beginning of the strike.[20] Approximately 8,000 telegraphers across the country joined in the action. Of these, somewhere between 300 and 1,000 women telegraphers joined in asking for equal pay for equal work. Women in Boston and Cleveland went on strike; they were joined by female telegraphers in Georgia, Florida, and North Carolina. In Baltimore, the seven women operators at the American Rapid Telegraph Company decided to strike on July 20, a day after the men walked out. In Chicago, 28 out of 30 women operators of Western Union joined the 120 men in walking out: "The men began filing toward the door, but before they reached it stopped and formed in line between the desks in the centre, and let the girls pass through, applauding them slightly." Eight of the twelve women working as Wheatstone operators in Chicago also joined the strike. Even though the Wheatstone telegraph was being abandoned in England, it had recently come into use at Western Union as a means of cutting labor costs.[21]

The strike slowly spread from the urban centers to the rural offices in all parts of the country. In Concord, North Carolina, Western Union closed the telegraph office and cut the wires leading to it after its operator there, Mary Ormand, joined the strike. Area newspapers championed her cause; the Charlotte, North Carolina, *Home and Democrat* reported that "the telegraph office at Concord has closed, as the W.U. could not persuade the brave Miss Ormand to be false to her womanhood." Mary Ormand first went home to South Carolina after losing her employment, but she was soon lured back to Concord by a promise of a job with a competing telegraph company, the Southern Telegraph Company, which evidently had no problems with her union membership. According to the Raleigh, North Carolina, *News and Observer*, "Two weeks ago the Western Union company cut its own wires and removed the instruments from the office in that town [Concord], thus depriving the citizens of the benefits of the wires, and all because Miss Mary Ormand, the operator, was a member of the Brotherhood and had joined the strike. The Western Union moved out of a warm berth and the Southern Telegraph Company, seeing its opportunity, hopped into the place vacated by the Western Union." Eventually Western Union reopened its office in Concord and brought in Alice F. Johnston, formerly the operator in Wadesboro, North Carolina, to run it. Johnston, a nonunion operator, had lost all of her belongings in a house fire shortly before in Wadesboro. The fire was apparently accidental and not related to the strike. The *Concord Register* noted the presence of the two competing companies: "Now we have two telegraph offices in town, and still we are— happy."[22]

Many railroad telegraphers joined in the strike and left the offices of the Baltimore and Ohio and Pennsylvania Railroads. When the B&O ordered the female operator at West Newton, Pennsylvania, to report to Pittsburgh and replace the striking operator there, she refused to do so and joined the strikers.

In Chicago, all twenty-two operators at the B&O office joined the strike. A *Tribune* reporter stopped by the B&O office to see how Mr. McCullough was faring:

> When a Tribune reporter called about 12 o'clock he [McCullough] was found trying to manipulate one of the instruments, but failing to get an operator at the other end, his success was rather limited. He was not left altogether alone, however. An envelope addresser stuck nobly to her post in the outer office, as also did a check-girl and office-boy. . . .
>
> The check-girl arranged her bangs, and fastened a tulip on her bosom, and gazed out on the windows across the street. The

office-boy pleasantly squirted tobacco-juice in unexpected directions, and skirmished back and forth in the corridor in a state of
gleeful excitement.

The *Tribune*'s reporter also called on R. C. Clowry, superintendent of
the Chicago Western Union office, and confidant of Eckert; Clowry would
one day succeed Eckert as president. Asked to compare the strike of 1870
with the present action, Clowry said, "The strike is a weak one compared to
that of 1870. Then all [managers] and men went out. In the present case,
however, no officer, no office-manager, no chief-operator, nor subchief has
left the office." When asked, "Are not female operators worth as much as
men?" Clowry answered, echoing Eckert's statement: "Other things being
equal the services of a man are preferable to those of a woman for many reasons. Women are not so reliable, they are more subject to indisposition, and
they cannot be called upon in emergencies for night duty."[23] Women were
said to be more frequently absent than men; the opposition to women
working nights, however, seemed to exist primarily in the minds of men, as
there were numerous cases of women volunteering for night work when circumstances made it appropriate.

More than in 1870, the women strikers were seen as representing an
agenda that was distinct from that of the men in some respects. Although
there were common interests, such as pay increases and improved working
conditions, there was also the notion of equality, both in pay and in representation in the labor movement. Newspapers noted that in the Brotherhood of Telegraphers, women served on committees and had an equal voice
in determining policies. Supporters of the women's suffrage movement
began to take notice of the union's support of equal rights. Lillie Devereux-
Blake, a prominent leader of the suffrage movement, praised the telegraphers' union for its support of equal pay for equal work: "You simply
demand what justice would at once concede, and if you succeed in securing
the establishment of this rule that the labor accomplished and not the sex of
the operator, shall regulate the amount of salary paid, you will have done
much towards making this the usage in other branches of business."[24]

Some female spokespersons emerged in this strike, including Minnie
Swan, a twenty-year-old first-class operator in New York who had worked
for both Western Union and brokerage offices and who quickly became

known as the leader of the women members of the Brotherhood of Telegraphers. Swan clearly captivated the reporters for the *New York World;* they described her as a "pretty brunette" with "large black eyes, which sparkle like
the diamonds in her ears." She had given up her $80 a month position with
a brokerage firm to lead the striking women operators "and share their for-
tunes, whatever they might be." Evidently a charismatic figure, Swan initiated the walkout of the female employees of Western Union by simply
walking into the operating rooms of the city department on the day of the
strike; all but twenty-one of the one hundred women working there got up
and followed her out of the building. Swan put her ability to give forceful
speeches to work for the Brotherhood; at a meeting in Caledonia Hall on
July 22, 1883, she was called to the platform to receive a bunch of lilies on
behalf of the striking women operators and responded, "These lilies are
symbolical of the purity of our motives; and though they may fade and
wither, our purity of motive and the bands that connect us will not wither.
Our Brotherhood will never die."[25]

On the evening of August 8, 1883, a benefit concert and ball were held
for the striking telegraphers in Madison Square Garden. Local regimental
bands volunteered to play for the event, which was attended by more than
three thousand people. At ten o'clock, as the ball began, a procession of
eight hundred telegraphers was led into the ballroom by master workman
John Mitchell, a New York strike leader, and Minnie Swan. A romance had
already begun between the two strike leaders; the following year they would
marry.[26]

By August, though, it was clear to the telegraphers that their cause was
lost. Western Union was still in business, despite their best efforts, and was
doing a reasonable job of getting messages out on time using nonunion
telegraphers. As in 1870, Western Union employed strikebreakers, both
men and women. Some were unemployed telegraphers or graduates of the
telegraphy colleges who had been unable to find work; others were students. Finally, on August 17, 1883, John Campbell declared the strike at an
end.[27]

The editor of the Wadesboro, North Carolina, *Anson Times* tersely
summarized the entire labor action in a single sentence: "The great telegraphic strike has ended, as far as the South is concerned, and the operators
have returned to work without having carried a single point." The strike of

1883 ended up repeating the mistakes of 1870. Like the earlier strike, it was completely unsuccessful; Western Union was simply too large and powerful. Hoped-for financial support from the Knights of Labor never materialized. When their strike funds were finally exhausted, the strikers capitulated and returned to work at the old wage levels.[28]

As in 1870, it was charged that the women operators were disproportionately victimized by the telegraph companies and denied reemployment after the strike. The women operators, by and large, had "stuck it out" to the bitter end, refusing to return to work as long as the Brotherhood was still on strike. After the strike, many refused to submit to the humiliation of begging for their old positions, including "Miss X," who was interviewed by a *New York World* reporter after a union meeting on August 18, 1883:

> "That smirking Mr. Dealy! He thinks he has broken us all up," said one of the young ladies after the meeting. "When he refused to take me back he rubbed his hands and said: 'Oh, I am sorry, Miss X, but really I can't use you. You must look elsewhere for employment.' He thought he crushed me, but he didn't. I turned round and said: 'Keep your position! I don't want it, and if you think you can deprive me of bread in that way you make a grand mistake—don't forget it!'"
>
> "What do you propose to do?"
>
> "I can say that I am expert with the needle and also write a fair hand, so I have more than one way to make an honest living."

William J. Dealy, manager of the operating room of the New York Western Union office, clearly took advantage of the situation to eliminate union sympathizers and other potential troublemakers from the ranks of telegraphers to be rehired. Another operator, interviewed by the same reporter, was clearly disillusioned by the outcome of the strike but just as unwilling to swallow her pride:

> "I will never touch a key again in my life," said another. "With the unsuccessful end of the strike my telegraphic days end also. I would rather cut off my right hand than humiliate myself by asking for my old place. Don't think that I could not get back if I

wished. During my employment in the main office I earned as
much as anybody in the service, but I never hope to go back. I can
do anything—wash, scrub, cook, sew or teach children. I will get
something to do, especially since I am willing to take a turn at anything."[29]

The September 7, 1883, issue of the *New York Times* carried a notice of a
fund being raised to aid the women strikers who had been denied reemployment: "A fund is being raised in behalf of the telegraphers engaged in the
recent rising against the Western Union Company who have since been refused employment. . . . Aid is especially asked for the ostracised female operators, who, after their hearty encouragement when the men showed
symptoms of yielding, have been persistently rejected when applying for
their old situations. These ladies, however, have no desire to live in idleness,
and ask those who sympathize with them to give them notice of any situations that can be obtained." Among those who contributed to the fund were
Thomas Edison, who, though possessed of a low opinion of women operators, had long sympathized with the desires of telegraph operators to improve their wages and working conditions, and the Cigarmakers' Union,
whose members were evidently willing to overlook some of the conde-
scending remarks made by the striking telegraphers and aid them in their
hour of need.[30]

Like the TPL, the Brotherhood of Telegraphers soon faded away, its
credibility destroyed by the unsuccessful strike. It withdrew from the
Knights of Labor in September 1883, charging that the Knights of Labor
had failed to support it financially during the strike; as John Mitchell remarked bitterly to a *New York World* reporter, "We were shamefully treated
by them and we will take decisive action. Don't be surprised if we desert the
order. They were the cause of our failure. We have always paid our assess-
ments regularly. When we take action in their case the public will know it.
Hereafter all our meetings will be as secret as before the strike."[31]

WOMEN OPERATORS AND THE WOMEN'S MOVEMENT

Although the strike failed to improve the status of women in the industry
or to reduce the inequity in pay, it did mark the beginning of an association

between the women's suffrage movement and the women operators, as supporters of suffrage took note of the Brotherhood of Telegraphers' demands for equal pay and equal representation. At the same time, female telegraphers began to realize that having the vote would greatly enhance their chances of improving working conditions and eliminating the inequity in pay through legislative action.

A similar alliance of interests would be formed in Norway in 1898, when women operators asked the Norwegian parliament (Storting) for an increase in pay. The Norwegian feminist union (*kvinnesaksforening*) sent a statement of support to the Storting. The women operators in Norway, however, did not have the support of the telegraphers' union (Telegraffunksjonaerenes Landsforening) of which they were members; the union, which had previously requested, and obtained, a pay increase for its male members, actually spoke against granting a pay raise for the female telegraphers, claiming that they were lacking in "technical skills, stamina, and independence." Finally, unable to obtain backing from the telegraphers' union, the women operators in Norway formed their own labor association in 1914, which immediately affiliated with the National Council of Norwegian Women (Norske Kvinders Nationalraad).[32]

THE ORDER OF RAILROAD TELEGRAPHERS

Railroad telegraphers, feeling that their interests had not been represented adequately by the Brotherhood, organized their own union in 1886. From then on, railroad telegraphers considered themselves occupationally distinct from the commercial telegraphers who worked for the large telegraph companies. The Order of Railway Telegraphers was founded in Cedar Rapids, Iowa, in 1886; it was the first union formed specifically to address the needs of railroad operators. (The name was changed to the Order of Rail*road* Telegraphers in 1891.) In fact, the ORT was ideologically closer to the conservative railroad unions, like the Brotherhood of Locomotive Engineers, than to the more militant commercial telegraphers. As part of the reaction to the abortive strikes of 1870 and 1883, it even had a clause in its articles of association that forbade strikes except under extreme conditions. By 1903, when Ma Kiley joined the ORT in Durango, Mexico, the union

boasted a membership of around twenty thousand, which constituted perhaps half of the railroad telegraphers in the United States. The ORT func-
tioned primarily as a benevolent organization that paid benefits to
out-of-work members and helped them to find work.

Although some local districts resisted admitting women, female railroad operators sought admission to the ORT in increasing numbers around
1900. By 1905, the ORT had at least one female district representative,
Katherine B. Davidson.[33]

Figure 21. ORT members, Columbus Grove, Ohio, 1907. From
Railroad Telegrapher, May 1907, 816. Reproduced from the Collections
of the Library of Congress.

THE COMMERCIAL TELEGRAPHERS' UNION AND
THE STRIKE OF 1907

Around the turn of the century, the commercial operators again began to form labor organizations. The Commercial Telegraphers' Union of America was formed in 1902 as an offshoot of the ORT for the specific purpose of organizing commercial operators. The following year, it merged with a similar group that had sprung up independently, the International Union of Commercial Telegraphers, and became a constituent union of the American Federation of Labor. Its stated goal was to gain membership among Western Union employees and eventually to represent Western Union telegraphers in negotiations with the telegraph company. By this time, there were only two large commercial telegraph companies, the Western Union and the Postal Telegraph Company, who together monopolized the telegraph business in the United States. The CTUA, under the leadership of President Samuel J. Small, had ten thousand members by 1904.[34]

Unlike earlier telegraphers' labor organizations, the CTUA's constitution contained language specifically excluding nonwhites, reflecting the tendency of unions of the age to focus on protecting jobs from inroads by foreigners and minorities. Although progressive elements in the union periodically tried to have the whites-only clause removed (finally succeeding around 1950), they were repeatedly voted down after being accused of fomenting "socialism."

While many railroad telegraphers may have shared in the CTUA's overt racism, the ORT had more tolerant admission policies, particularly regarding operators with Hispanic surnames. This was in part owing to the strength of the ORT in Mexico, where it had many large and active districts; photographs of Mexican ORT officials were featured prominently in the ORT's journal, the *Railroad Telegrapher*.

Background of the 1907 Strike

By the beginning of 1907, discontent was spreading among telegraphers and it was clear that a strike was possible. The issues were similar to those that precipitated previous strikes—low pay, long working hours, bad working conditions. In Chicago in 1907, pay rates for telegraphers ranged from

$25.00 to $82.50 a month. Working hours were typically ten hours a day, six days a week, with frequent Sunday work as well, in operating rooms that were often poorly lighted and ventilated. Three-quarters of women operators made less than $45.00 a month. Pay rates had not increased since the 1880s, while buying power had gone down. Wages were actually numerically lower than they had been during the 1860s.[35]

Another issue that was causing discontent among operators was the cost of providing a typewriter. In many offices, operators had to provide typewriters at their own expense. Referring to telegraphers in her 1909 study of Pittsburgh workers, *Women and the Trades*, Elizabeth Butler noted that when typewriters were first introduced into the telegraph office, bonuses were offered to telegraphers who would learn to use them, and their use quickly became widespread. Operators were required to purchase their own typewriters, however, which constituted a considerable initial investment.

Butler also noted that the sliding scale was being used primarily to replace men with women at a lower rate of pay in 1907:

But important as these points were, abolition of the sliding scale was the cardinal demand. The grievance referred to as the "sliding scale" was the outcome of alleged differences in the work done by men and by women, and of resultant unfair discrimination. . . .

Although the work might tell on women sooner than on men, and although they might in some cases be less efficient than men, they were yet sufficiently capable to supersede men at a lower rate of pay. They were lending themselves to a scheme for cutting wages.[36]

Union activities among its operators had begun to come to the attention of Western Union by early 1907. The company dismissed two female CTUA members in New York City in April 1907; Sophie Annaker had complained of poor lighting, and Camilla Powers was told by Western Union chief clerk T. Brennan, "You are a union agitator and we don't want you." Powers was well-known and respected by other operators; she was a second-generation telegrapher, the daughter of Samuel L. Welp, who had operated in New York and Chicago. In May, the CTUA drew up a list of grievances and presented it to Western Union president R. C. Clowry, charging that the company had failed to live up to its earlier promise of a 10 percent raise and was summarily dismissing employees for wearing union buttons.[37]

In June 1907, operators in San Francisco requested a temporary 25 percent pay increase to cover additional living expenses caused by the San Francisco earthquake. The General Executive Board of the CTUA met in New York and, finding that the San Francisco operators had just cause for their request, authorized calling of a strike if Western Union turned down the request; they sent President Small to San Francisco to investigate the situation. Small left New York on June 14 with the authority of the Executive Board to declare a strike, but he intended to avoid one if at all possible. Meanwhile, in New York, Sylvester J. Konenkamp, a member of the General Executive Board, began to negotiate with Commissioner Charles P. Neill of the U.S. Labor Bureau, who was also talking to Clowry of Western Union. Neill was working to have identical agreements signed between the CTUA and Neill and between Western Union and Neill to the effect that the 10 percent raise promised earlier by Western Union would be put into effect; the separate agreements approach was used to avoid having Western Union negotiate directly with the union, which the company did not recognize.[38]

In San Francisco, Small addressed the issue of pay for women operators. He noted that while male operators in San Francisco made an average of $70 a month, "There are many female operators working for the company today who only receive $40 a month, and are compelled to pay $2 to $5 a month for a typewriter, as without a machine they cannot hold a job."[39] Western Union officials in San Francisco rejected the 25 percent wage increase demand. Small in San Francisco either did not know of the New York agreement or did not consider it valid. The CTUA did not use the telegraph to communicate with Small in San Francisco for fear the message would be intercepted, and cross-country long-distance telephony did not yet exist. He called for a strike on June 21, after being given authority to do so by a vote of the San Francisco union local. About 200 telegraphers, 150 Western Union employees, and 50 employees of the Postal Telegraph, joined the strike at Oakland. The *San Francisco Chronicle* reported that as in earlier strikes, women operators who joined the cause were "received with hand-clapping by the male operators who were standing in the nave of the long building." The newspaper also intimated that the strike was no local "wildcat" action but part of a well-orchestrated nationwide strategy to disrupt telegraph service and demand concessions from Western Union: "Although the union leaders insist that the strike here is merely local, there is good ground to believe that it is part of a well-planned strategy which has been developed while the trouble was brewing in every city from here to New York."[40]

Negotiations between the CTUA, Western Union, and Commissioner Neill began on July 12. Neill was soon replaced by his assistant, Ethelbert Stewart. A preliminary agreement was reached on July 19 in which Western Union agreed to rehire the operators who had gone on strike on June 21 and to discuss salary increases after full resumption of work. The union local in San Francisco voted to accept these terms, and at first it appeared that the threat of a nationwide strike had been averted.[41]

Events that occurred shortly thereafter, however, dashed hopes of a negotiated settlement. During the San Francisco strike, Sadie Nichols, a nonunion beginning operator at San Francisco, was put on the wire to Los Angeles, a position normally reserved for first-class operators, as a reward for not joining the strike. The union operators in Los Angeles learned that she had been a "scab" during the strike and began to harass her. Operator Paddy Ryan in Los Angeles refused to work with Nichols and told her over the line that "the place for her to live was in a notorious bawdy house in San Francisco." She in turn called Ryan a "liar." She reported Ryan to management on July 23, and he was fired after he was taped and it was proven that her accusations were correct. (Being taped meant that management put a recording telegraph on an operator's line to determine exactly what was being sent and received.) On August 7, operators at Los Angeles went on strike to show their support of Ryan.[42]

Women operators had the choice of viewing the incident in California as either a women's or a labor issue. Most seemed to side with the union and to blame Western Union for placing Nichols in a position for which she was not qualified. In Chicago, operators went on strike on August 9, refusing to work with the nonunion operator in Los Angeles who had replaced Ryan. At this point, the CTUA, presented with a fait accompli by the rank and file, had no choice but to authorize the strike on August 15. Meanwhile, the ORT had ordered its members to refuse Western Union traffic on August

11; on the same day, Associated Press operators struck for higher pay. All told, between ten thousand and fifteen thousand operators struck across the country during the summer of 1907.[43]

The CTUA and the WTUL

Because of its central location and economic prominence, Chicago became the center of the strike action in August 1907. The women strikers were promised "financial and moral support" in Chicago by Jane Addams, Ellen M. Henrotin, Mary McDowell, and Margaret Dreier Robins, the leaders of the National Women's Trade Union League, in an article that appeared in the *Chicago Tribune* for August 11, 1907. Cora Talmadge and Delia Reardon of the Chicago local of the CTUA were appointed to confer with the NWTUL.

The NWTUL (or WTUL, as it was commonly abbreviated) had recently moved its center of operations from New York to Chicago, the residence of most of its leaders at the time. Jane Addams had long been associated with Hull House, a Chicago settlement house that was involved with labor issues, as well as improving living conditions for immigrants and low-wage workers. She had been appointed vice-president of the National Women's Trade Union League in 1903. Ellen M. Henrotin, a Chicago reformer, was elected national president of the WTUL in 1905. Margaret Dreier Robins was a reformer from Brooklyn, New York. Daughter of a wealthy merchant, she had become involved in progressive politics in 1903 and took a special interest in improving the working conditions of immigrant women; she became convinced that trade unionism was the best vehicle for achieving this objective. Margaret Dreier Robins was elected national president of the WTUL in 1907 and served in that capacity until 1922.[44]

A liaison between the WTUL and the CTUA had already begun before the strike. Members of CTUA Local 16 met with members of the WTUL in New York City on July 14, 1907, to explore issues of common interest and to work toward the mutual goal of organizing the five million wage-earning women in the country into a single labor movement. Local 16, headquartered in New York City, was a large local with over two thousand members, a high percentage of whom were women; it was commonly referred to in the *Commercial Telegraphers' Journal* (the official journal of the CTUA) as "Big 16." Speakers from the WTUL included Rose Pastor Stokes, Rose Schneiderman, and Harriet Stanton Blatch. Stokes was a former cigar wrapper whose unlikely odyssey from the slums of Cleveland to marriage into one of America's wealthiest families made her one of the best-known women of the age. Schneiderman, a former capmaker, played a large role in organizing women in the garment trades. Blatch was the daughter of women's rights advocate Elizabeth Cady Stanton and mother of Nora Blatch, who graduated from Cornell University with a degree in engineering and was briefly married to radio pioneer Lee De Forest. Among telegraphers attending were Mazie Lee Cook, said to be the highest salaried woman telegrapher in the country, Hilda Svenson, a telegrapher who represented Postal Telegraph employees, and Florence Worthington, who represented Western Union operators.

In her speech, Rose Pastor Stokes pointed out that the primary focus of

Figure 22. *Left,* Mary Dreier; *right,* Margaret Dreier Robins. Courtesy Margaret Dreier Robins Collection, Department of Area and Special Studies Collections, University of Florida.

the meeting was to include the telegraphers in an effort to organize the five million wage-earning women in the United States, an effort she described as "one of the most important steps taken in the labor movement." She stressed that the goal of the WTUL was not to set up separate unions for women workers but to encourage men and women to work together cooperatively to improve wages and working conditions for all: "The women want to cooperate with the men and become a part of their organization." Stokes also discussed the relationship of the union movement to the women's suffrage movement and the importance of involving working-class women: "This is one of the most important movements for women ever launched. In time I believe it will lead directly to general female suffrage, because it will bring women to the realization of their inability to better the conditions under which they work to any great extent, unless they have the vote, and the political power that goes with it. Agitation for female suffrage, when backed and obtained by this class of women, will be far more effective than by the wealthier class of club women who have heretofore been behind it."

Figure 23. Hilda Svenson, Postal Telegraph strike leader, New York, 1907. From *Commercial Telegraphers' Journal*, October 1907, 1076. Reproduced from the Collections of the Library of Congress.

Harriet Stanton Blatch agreed that women workers could be more effective than upper-class women in getting the vote for women: "It's the factory girls, and not the members of women's clubs, who are going to get the suffrage," she declared, "and it's coming soon. Such a force as this will bring it about in no time. These girls are fighting practical business propositions every day; they know they must have a vote if they win out."

Rose Schneiderman noted that while men sometimes held union meetings in saloons, women should not use that as an excuse not to attend, declaring that "I'd go to a meeting if I had to go through five saloons to get there. Don't sit back and let the men do your work for you. Take your own stand, fight your own battles, and don't be afraid."

Mazie Lee Cook, reputedly the highest salaried woman telegrapher in the country at a pay rate of $42 per week, spoke about the importance of working with men in the labor movement. She asserted that "men generally object to women working for less than themselves. They are anxious to elevate the women to the equal pay standard." Cook spoke of her own experience of going to work for Western Union in New York at the age of sixteen for $12.50 per week, while her male partner on the wire was paid $17.50 for the same work. She spoke in favor of equal pay for equal work and stated that women should not be required to work first-class wires without getting first-class pay.[45]

The WTUL continued to work in support of the CTUA in New York after the strike was declared. On August 25, Rose Pastor Stokes delivered a speech to the striking telegraphers at Everett Hall, CTUA headquarters, citing the unfairness of companies in requiring telegraphers to furnish their own typewriters. Beginning her speech with "Friends and comrades, for I understand there are some Socialists among you," Stokes continued: "An eight-hour day must be believed in by all fair-minded people both among the workers and the employers. As long as the company believes you can buy your own typewriters it will insist on your buying them and take the profits for your work. It was so with us in the cigar trade and it is so in all trades in which hand tools are used."[46]

In Chicago, the striking telegraphers also received the support of the Australian reformer Alice Henry, office secretary of the Illinois branch of the WTUL, who spoke to a meeting of strikers at Brand Hall, the CTUA union hall in Chicago:

"I have been much struck," she said, "during the last few years
with the wonderful unanimity of the American unions on the
question of equal pay for equal work. This touches everybody.
"I bid you 'stick'! You are going to benefit the world."[47]

Women Strike Leaders

Women operators gained visibility as strike leaders when events thrust them into the forefront of the labor action. Louise Forcey, a Chicago CTUA member, earned a write-up in the *Chicago Tribune* for her enthusiastic efforts to rally telegraphers to the union cause when she encouraged operators to walk out of the offices of the Postal Telegraph Company on August 9.

The strike began for the Chicago employees of the Postal Telegraph at ten minutes past six on August 9, when M. J. Paulson, president of the union local, and E. M. Moore, chairman of the local executive board of the union, entered the Postal operating rooms, where 150 operators, 120 men and 30 women, were busily working. Paulson blew a whistle he was carrying, and, according to the *Tribune*, "the whole 150 of them jumped from their chairs as if propelled by steel springs." At this point, the *Tribune* continued, Forcey took charge and ordered the strikers to remove the "tabs" that indicated the call signs of other stations on the line, making it difficult for strikebreakers to operate; she became, in the words of the *Tribune*'s reporter, the "Joan of Arc of the strike movement." Forcey then turned her attention to some operators who had refused to join the strike:

> Over in the east end of the room the men had massed together in concentric rings around half a dozen operators who had remained in their seats. Miss [*sic*] Forcey broke through the rings and energetically pleaded individually with the recalcitrants to "come out." The strikers cheered every sentence she uttered in her

Figure 24. *Right*, Mrs. Louise H. Forcey, Postal strike leader, and fellow strikers "turn their hands and sweetest smiles to new occupations." From *Chicago Tribune*, August 15, 1907.

fervid exhortations. Once in a while she became hopping mad when her words appeared to have no effect and spoke her thoughts right out in meeting. Other girls did their best to aid her by loudly calling out the full name of every man who had not signified his intention to leave.[48]

Women also joined the picket lines during the 1907 strike. A female picket appeared on Lasalle Street in Chicago, where both Western Union and Postal Telegraph headquarters were located; she was described by the *Tribune* as a "demure little creature who presumed to guard the doors of both companies from the unwelcome presence of feminine strike breakers." Giving her name only as "Miss Millner," she told the press: "We women are in the strike to win. . . . The men are doing a great deal for us when they insist that we have equal pay with them and when they declare that they will stay out until the companies grant that demand. It is little enough for us to help them as much as we can."[49]

In other parts of the country as well, women entered leadership roles as women's issues came to the forefront. In St. Louis, Eva E. Tracey, who had been pictured in the June issue of the *Commercial Telegraphers' Journal* clowning in a skit at the Grand Musical Entertainment and Dance held by the St. Louis CTUA Local No. 3, appeared in a more somber role in the September issue of the same journal as a member of the union's Executive Board.[50]

Hilda Svenson became a leader of the Postal Telegraph strikers in New York City. She was interviewed by the *New York Journal*, which referred to her "command of 345 young lady pickets." Nellie E. Pearl, her counterpart for Western Union, was interviewed by the *New York World*, which featured pictures of her and a description of a day on the picket line.[51]

Media Coverage

Although the strike was enthusiastically covered by reporters for the *Chicago Tribune*, its reporting was often superficial, focusing on "glamour shots" of attractive women and "human interest" stories; there was little mention initially of women's strike issues. *Tribune* reporters frequently portrayed women strikers as hysterical and inexperienced; an account of a union meeting at CTUA headquarters stated, "A strike was a new thing for the women and they probably were excusable for being wrought to a pitch of hysterical excitement by the strong words and arguments that were advanced by the more experienced men."[52]

The *Tribune* reporters turned their strike coverage into a "beauty contest," concentrating on photos of Lillian Sullivan's "striking smile" and Ellen Forsman, "the prettiest of the striking telegraphers." Women strikers themselves objected to not being taken seriously; when a group of them showed up for a meeting at the union hall, they were asked by male union officials to go outside and pose for pictures, as was mentioned in the August 13 issue of the *Chicago Tribune:*

> Most of the girls who have joined the men operators in the telegraph strike find it difficult to have themselves taken seriously. Fully 100 of them appeared at Brand's Hall shortly after noon yesterday and waited patiently in the gloomy corridors for the mass meeting at 4 o'clock. But when the subject of the girls was broached to the men they glanced at the picture hats and fluffy summer gowns and smiled.
>
> "Well, it's not our fault," said Miss Kate Watkins indignantly, when an outsider intimated that the young women seemed to be having a beautiful time. "What do you think was the first thing we heard when we reached the hall this afternoon?"
>
> "Harry Likes [chairman of the strikers' grievance committee] was up there in front and when he saw us he said, 'O girls, there's a photographer outside who wants to get some pictures of you. Go on out, won't you, all of you? You know, the more pictures they take of you, the more it helps the cause.'"
>
> "So we all went out and left the boys in there to do their talking alone. Humph! I s'pose they think that's all we're good for!"[53]

Letter to Helen Gould

The August 18 issue of the *Chicago Tribune* included a copy of a letter from the women strikers addressed to Helen Gould, which brought up what the *Tribune* referred to as the "moral issue" raised by the striking telegraphers.

Gould was the daughter of Jay Gould, the Gilded Age baron who had once counted Western Union as one of his fiefdoms. Helen Miller Gould Shepard (1868–1938) had a reputation as a philanthropist who was sensitive to women's issues; she was also a major shareholder in Western Union.

The letter began by stating that the upper management and board of directors of Western Union had been deceived about the true state of affairs in the company. The women strikers termed their action a "revolt" rather than a strike and then proceeded to list the discriminatory actions that had precipitated the revolt.

The first item mentioned by the women was the condition of their bathrooms and private facilities: "Our withdrawing rooms and conveniences are a disgrace to humanity." The strikers alleged that the health department had ordered the company to make improvements, but the recommendations had been ignored. They also noted that they were required

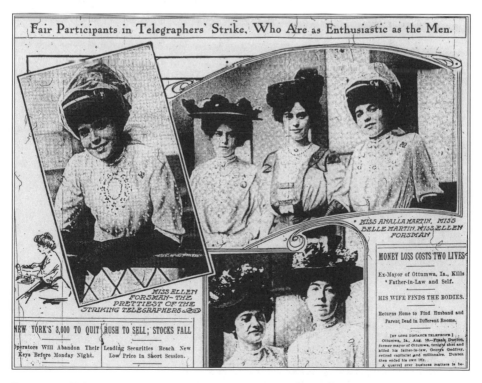

Figure 25. *Chicago Tribune* coverage of the 1907 strike. From *Chicago Tribune*, August 11, 1907.

to enter the Western Union building by a rear entrance, which made it necessary to pass "through an alley lined with saloons and rubbish." The signers of the letter said that the "employment of extremely young girls" by the company, whose duties brought them "in personal contact with men of all classes, and absolute lack of discipline among them," created the possibility of sexual abuse; they went on to point out that the company had placed men of questionable morals in positions of responsibility over women and children. The letter closed with a personal appeal to Helen Gould: "We beseech you as an American woman to take up this cause and get us the justice which we can never get through characters concerning whom, upon investigation, you will find we have spoken the absolute truth."[54]

The issue of unequal pay was not mentioned; rather, the women operators concentrated on issues specific to the employment of women that they felt might gain a sympathetic hearing from Gould. The issue of women employees being forced to use a rear entrance had been brought up a year earlier, in 1906, when telephone operators had struck in Chicago; the WTUL had supported the telephone operators in that strike as well.[55]

Margaret Dreier Robins enlisted the help of family members in support of the telegraphers. She wrote to her sister Mary Dreier in New York on September 6, 1907, about the possibility of having a third sister, Katherine, a Western Union stockholder, write a letter to Western Union management. She felt that Katherine would be seen as a more impartial voice than either Margaret or Mary, both of whom were closely connected with the WTUL. She suggested that Katherine could point out that the company was losing the confidence of the public by its refusal to accept arbitration; in addition, "if she will say that women doing the same work as men are entitled to the same wage and that typewriters ought to be furnished by the company, and then call attention to the evils of the commission office which are not even touched upon by the strikers, I think she will make a strong letter."[56]

Robins was concerned about the practice of placing women operators in commission offices, where their low pay allegedly left them susceptible to the wiles of the "sporting element" that frequented such places. Robins also wrote directly to Helen Gould regarding the telegraphers' cause; in her letter, she raised the issue of pay. She discussed the letter to Gould in a letter to her sister Mary, dated September 12, 1907:

We are very busy with the telegraphers. We are sending out letters to the allies of the league asking for twenty-five cents a week for four weeks to be used for the benefit of the telegraphers. I am going to try to get the league to publish a printed letter stating some of the conditions under which women work. In my letter to Helen Gould, I emphasized these facts.

The physical demand of the work is such that that alone entitles the operators to a short working day. The nervous strain often causes a paralysis of the right hand as well as a complete nervous collapse and occasionally it produces insanity.

According to the government figures, thirty-nine dollars a month is the average wage paid, and this is less than ten dollars a week—a sum on which it is quite impossible for any self-respecting woman to live in our large cities.[57]

The appeals to Helen Gould and the management of Western Union, however, had little effect. The strike became a waiting game in mid-August, when it became apparent that the union did not have the money to carry out a protracted strike. President Small made extravagant claims about raising a $2 million strike fund, but the hoped-for support from the American Federation of Labor and other sympathetic groups never arrived; members of the union's executive board began to question his sometimes erratic and impulsive management of the strike. Telegraphers began to desert the strike, some accepting railroad positions and some going back to Western Union and the Postal Telegraph. The economy, which had faltered several times during the year, began to lurch ominously toward a depression in the autumn as stock speculation sent several prominent banks into receivership and anxious depositors lined up to withdraw their funds. Small sent out a telegram on October 12, calling for an end to the strike; the CTUA Executive Board, angered that he had acted unilaterally, suspended him.[58]

In mid-October 1907, Margaret Dreier Robins went to Rockford, Illinois, with Delia Reardon of the Chicago CTUA, to address the annual meeting of the Illinois State Federation of Labor, which met in Rockford on October 15–18. The Federation of Labor devoted an entire day to women in the labor movement; according to Robins, they received a "tremendous reception." By this time, however, the strike was over, for all practical purposes, and the telegraphers were reduced to asking for contributions from other labor unions to help those who could not find jobs. Delia Reardon delivered a speech to the State Federation of Labor about the telegraphers' strike and the working conditions they were protesting; she mentioned the requirement for telegraphers to furnish their own typewriters, the inequity in pay between men and women, and the refusal of the strikers to accept inferior positions after the cessation of the strike action. According to the *Rockford Daily Register-Gazette*, "Her recital of the conditions against which the telegraphers struck received a sympathetic hearing from the delegates." But their plea for donations yielded a pledge of only $133.20 from the delegates.[59]

End of Strike—Effect on Women

The strike was officially called off on November 9, 1907. Like the previous failed uprisings on the part of the telegraphers, the strike of 1907 failed for four reasons: lack of an adequate strike fund, lack of advance planning or forethought, failure to consider the financial strength of the telegraph companies, and an abundant supply of nonunion operators to take over the jobs of the strikers. Although telegraphers would strike again, on a smaller scale, in 1919, 1929, 1946, and 1948, the 1907 strike was the last time that a nationwide shutdown of the telegraphic system was used in an attempt to win concessions from the telegraph companies.[60]

Another factor, especially in the strikes of 1883 and 1907, was the failure of other organized labor groups to come to the assistance of the telegraphers. This was in part because the Knights of Labor and the American Federation of Labor were perennially strapped for cash; but it was also owing to what the more solidly working-class unions perceived as the "elitism" of the telegraphers. To pipefitters and boilermakers, the telegraphers were "kid-gloved laborers" who had lost touch with their working-class roots, as noted in an article about the strike that appeared in the *Blacksmith's Journal* for October 1907: "Telegraphers are no longer restrained by the feeling of deference akin to that of personal servants; a feeling that probably militated against them and made the former strike against the telegraph company more easily defeated than it would have been if the operators could have known, or be impressed, that they were wage workers and subject to all that other wage workers have to bear, and that it is only through organization that they can preserve their rights and secure the conditions to which they are entitled."[61]

Although the strike did not achieve its objectives, women operators in particular did obtain some improvements in their working conditions in its aftermath. Working hours for women were shortened as a result of congressional inquiry. Although no wage concessions were granted by Western Union, the Postal Telegraph gave employees a 10 percent increase in wages, bringing some women operators into wage equality with men. The percentage of women employed by the telegraph companies increased in 1908, as large numbers of women were hired to replace men who were not rehired after the strike.[62]

Ironically, working conditions improved in 1909, not because of the protests of the strikers but because of reforms brought about by the merger of AT&T and Western Union. Theodore Vail, president of AT&T, brought about many changes in the management of Western Union, including improvements in working conditions and facilities. Even the *Commercial Telegraphers' Journal* hailed Vail as a man of "broad and liberal views."[63]

For some women participants, the strike brought new visibility and new roles. Hilda Svenson became a professional organizer for the WTUL. Mary Macaulay used the experience she gained as vice-president of Buffalo CTUA Local 41 to become international vice-president of the CTUA in 1919. And in Atlanta, Georgia, Ola Delight Smith became a journalist for a local labor newspaper, the *Journal of Labor*, after being blacklisted by Western Union for her union activities.[64]

For Ma Kiley, sticking to her principles cost her a job during the strike of 1907, "the only time in my life," as she recalled, that she was actually fired. Working as a railroad operator for the Chicago, Rock Island, and Pacific line in Waurika, Oklahoma, she was obligated as an ORT member to refuse to send any Western Union messages over the line during the strike. When a group of men first tried to bribe her and then to coerce her into sending a telegram, she replied, "I AM NOT SENDING THAT MESSAGE!" One of the men reported her to railroad authorities, and she was summoned to railroad headquarters in Fort Worth and fired.[65]

THE STRIKE OF 1919

After the United States entered World War I, President Woodrow Wilson nationalized the telegraph and telephone services and placed them under the direction of Postmaster General Albert S. Burleson. Burleson was seen by many telegraphers as favoring the management of Western Union; in 1918, under his administration, Western Union was allowed to form a company union, called the Association of Western Union Employees, and to attempt to bar its employees from joining the CTUA. When Western Union discharged several operators for belonging to the CTUA, the union went on strike on June 11, 1919, asking for higher wages and equal pay for women. By this time, the CTUA had only about thirty-five hundred members; turnout at the picket lines was small. Reflecting the increased employment of women in the telegraph industry, however, approximately 30 percent of the pickets were female. Pickets from the Women's Trade Union League joined the telegraphers; Rose Schneiderman, now a leading organ-

Figure 26. Ola Delight Smith, labor journalist. From *Railroad Telegrapher,* May 1911, 1018. Reproduced from the Collections of the Library of Congress.

izer for the WTUL and a prominent spokesperson for suffrage, addressed the striking telegraphers.[66]

Women pickets were no longer looked upon as curiosities. Confrontations between strikebreakers and pickets ultimately resulted in arrests. In Oklahoma City, Oklahoma, three striking multiplex operators, Mrs. A. R. Payne, Ethel Osborne, and Myrtle Dever, were arrested and charged with "coercing and intimidating" Western Union employees.[67]

Like its predecessors, the strike of 1919 ended with capitulation on the part of the striking telegraphers. The president of the CTUA, S. J. Konenkamp, resigned; this led to the election of a new slate of officers, including Mary J. Macaulay as international vice-president, the first woman telegrapher to hold a national elective office in a union. One of her first actions was to set up a defense fund to aid the strikers who had been arrested in Oklahoma City. Her efforts proved successful in early 1920 when all of the federal indictments brought against the strikers were dropped.[68]

Macaulay served as vice-president until 1921. She developed a reputation for quiet diplomacy, and the *Commercial Telegraphers' Journal* lauded her negotiating skills, noting that "Sister Mary has on more than one occasion proven herself a real anchor when parliamentarian ability has been necessary to smoothly iron out convention difficulties."[69]

THE 1919 STRIKE IN AUSTRIA

Strikes occurred in Europe in the post–World War I era as well, as autocratic regimes were replaced by more democratic governments and women began to demand equality in the workplace. In January 1919, Austrian telegraph and postal workers went on strike to protest unequal pay and working conditions for male and female employees. Before the strike, women telegraphers were required to leave the service if they married; women received lower pay than men and were not allowed to become supervisors. As a result of the strike, a comprehensive policy of equal rights was enforced in the post and telegraphic administration; men and women were placed on an equal pay scale, and women were appointed to managerial positions for the first time. Women who had left the telegraph service to marry were allowed to reenter the service.[70]

THE DECLINE OF THE CTUA

Unlike its predecessors, the TPL and the Brotherhood of Telegraphers, the CTUA managed to survive two failed strikes, but it would be many years before the union fully regained the role it had played before 1907. After the strike of 1919, the CTUA went into a long period of decline. The telegraphers and their unions did not share in the prosperity of the early 1920s; the CTUA entered the decade with only about two thousand members and still had approximately the same membership fifteen years later, in 1935, after the Great Depression and increasing mechanization had taken their toll on the ranks of telegraphers. Automatic and multiplex operators formed an increasing proportion of the membership as the telegraph companies replaced Morse instruments with Teletypes. Women, however, were increasingly visible as officers of union locals; Anna Fallon, secretary-treasurer of the District of Columbia CTUA District, Division 55, was featured on the cover of the *Commercial Telegraphers' Journal* for March 1937, signing a contract between the CTUA and officials of the Postal Telegraph Company.[71]

Figure 27. Anna Fallon signing agreement. From *Commercial Telegraphers' Journal*, March 1937, cover. Reproduced from the Collections of the Library of Congress.

In 1937, a rival union, the American Radio Telegraphers Association (ARTA) was chartered by the Congress of Industrial Organizations; its predominately male membership initially consisted of radio operators. The ARTA's charter was soon expanded to include land line telegraphers, and it changed its name to the American Communications Association (ACA). The ACA began to concentrate its efforts on organizing the Postal Telegraph Company and embarked on a campaign to draw membership away from the CTUA. The CTUA responded by accusing the ACA of being "infiltrated by Communists" and began an initiative to retain the support of the Postal telegraphers.

How to deal with the ACA was one of the primary issues on the agenda when the CTUA held its seventeenth Regular Convention in September 1937. For the first time since 1921, when Mary Macaulay had been international vice-president, there were two women delegates to the convention; Irene McCallie from Indianapolis, Indiana, and Celeste O'Grady from Kansas City, both representing the Postal Telegraph Division 55. Both were appointed to the Committee on Resolutions, which issued a proclamation under their signatures warning of the dangers of allowing the ACA to gain control of the Postal telegraphers: "It is the opinion of the Postal Delegates to this Convention that it is absolutely necessary that certain measures be taken if our objective of organizing the Postal is to be accomplished; further, that if such measures are not adopted we will be pushed from the field and the CIO will become the dominant factor in the Communications field."[72]

Their warning, however, came too late; in 1939, in spite of the efforts of the CTUA, the ACA signed a closed shop agreement with the Postal Telegraph Company, effectively shutting out the CTUA.

Having secured an easy victory over the Postal Telegraph, the ACA then turned its attention to Western Union. The National Labor Relations Board was persuaded to investigate the Association of Western Union Employees (AWUE), the company-sponsored union, which it found guilty of unfair labor practices. Western Union was required to disfranchise the AWUE and to hold an election in which employees could freely choose which union would represent them in negotiations with management.

Seeing the opportunity to achieve its main strategic goal, the CTUA was finally roused from its long slumber. The nearly dormant union began to campaign vigorously for the right to represent Western Union's operators, emphasizing its "American" roots and exploiting the charges that the ACA was dominated by Communists. Irene McCallie, now a full-time organizer for the CTUA, launched a campaign to organize Western Union operators in Washington. As a result of these efforts, the CTUA won the election and finally achieved its goal of representing the Western Union telegraphers. When Western Union and the Postal Telegraph were merged in 1943, a second election was held to determine union representation, and again the CTUA won in every district except New York City, which remained a stronghold of the more militant ACA. By 1944, the once dormant CTUA boasted an invigorated membership of twenty thousand members.

The ACA maintained its dominance of New York City in the post–World War II era, calling strikes against Western Union in 1946, when seven thousand New York telegraphers walked out, and against the cable companies in 1948, when approximately two thousand operators walked out of Mackay Radio, Commercial Cables, and the Western Union Cable Company. No significant women's issues emerged in the postwar era, and few women took part in the ACA strikes. Issues centered around working hours and pay rates, which were said to be among the "lowest for any major industry," reflecting the diminished importance of the telegraph business.[73]

The ACA was expelled from the CIO in 1950, during the anti-Communist fervor of the McCarthy era, and ceased to exist as a union. The CTUA, however, continued to represent diminishing numbers of telegraphers into the 1960s; in 1968, the union changed its name to the United Telegraph Workers.

The Order of Railroad Telegraphers experienced a long decline as well as its membership was diminished by the replacement of telegraphers with computerized train control. In 1965, the ORT changed its name to the Transportation Communications Employees Union, and in 1969 the railroad telegraphers' union was merged into the Transportation-Communications International Union.[74]

Many women telegraphers in the United States were dedicated union members who formed long-lasting relationships with the telegraphers' labor organizations and allied with and shared agendas with their male counterparts. One simple reason for this was that it was in their interest to do so.

The status of women in the telegraphers' labor movement can be viewed as a complex triangulation of interests. For the women themselves, telegraphy was clean, skilled work that paid more than most occupations open to women. For the telegraph companies, the entry of women into the profession meant reducing labor costs and the eventual deskilling of the work as the Teletype was introduced. For the labor unions, admitting women as full members and including women's issues in the agenda improved their bargaining position with the companies and reduced the likelihood that union members would compete with nonunion women for jobs. And, finally, for the women operators, membership in the union gave them representation and backing in their demands for equal pay and improved working conditions.

While the enthusiastic participation of women in the telegraphers' labor movement can be explained in these purely rationalistic terms, family and kinship ties probably played a role as well. As Carole Turbin notes in her essay "Reconceptualizing Family, Work, and Labor Organizing: Working Women in Troy, 1860–1890," the tendency of the working-class Irish to "network" among family and friends probably played a large role in the recruitment of union members among the predominantly Irish telegraphers. Combining ties of kinship and community with shared workplace concerns helped to create a strong basis for union organization.

The participation of women operators in the telegraphers' labor movement also led to closer ties with other groups working in support of women's rights. The connection with the women's suffrage movement in the United States dated from the strike of 1883, when Lillie Devereux-Blake praised the Brotherhood of Telegraphers for its support of equal rights and equal pay. The interaction between the WTUL and the CTUA during the strike of 1907 also focused on suffrage as a means of improving the status of women workers. As Rose Pastor Stokes and Harriet Blatch observed in 1907, obtaining the vote for women, and the political power that accompanied it, would give women workers a powerful tool in their struggle to achieve equality in the workplace. The election of Mary Macaulay, a longtime supporter of women's suffrage, to the position of CTUA vice-president in 1919 symbolized the commonality of interests between the women telegraphers and the suffrage movement.

Ironically, however, the achievement of full suffrage for women with the passage and ratification of the Nineteenth Amendment in 1920 also marked the end of an era of women's activism in the telegraphers' labor movement. The discussion of women's issues, which had been a prominent feature of the telegraphers' journals since the 1860s, largely disappeared in the 1920s. Although women continued to play a large role in union activities and to hold elective offices in many union locals, no women participated in union activities at the national level or served as delegates to the national union conventions until the mid-1930s.

Viewed in retrospect, the struggle of the telegraphers to raise their wages and maintain workplace autonomy through strikes and labor actions can only be seen as quixotic and an ultimate failure. Their strategy placed them on a collision course with an industry that was committed to the twin goals of automation and reduced labor costs, and the abundance of non-union telegraphers willing to work for lower wages deprived them of the most important tool available to unions—the ability to withhold their labor in order to bargain with their employers.

Conclusions

A S Melodie Andrews notes in her essay on women in the early American telegraph industry, the work of women in the telegraph office has been much neglected by labor and social historians. One reason for this lack is that labor historians have tended to focus on the introduction of women into gendered roles in the workplace as a result of industrialization, which largely took place after 1870, whereas women entered the telegraph industry much earlier and participated in its preindustrial phase.

In fact, the telegraph itself played a major role in the industrialization of business and work in the nineteenth century. The telegraph made the large industrial corporation possible by enabling buyers and sellers to conduct transactions quickly and efficiently over vast distances; it contributed to the development of large-scale capitalism and the market-driven economy by enabling stock and commodities prices to be quickly transmitted from one end of the country and, eventually, the world, to the other.

The industrialization of the economy brought about in part by the introduction of the telegraph led to social change, including the formation of a large middle class, which came to include telegraphers themselves. The values of the new middle class included self-improvement and upward social mobility; for women in particular, membership in the new middle class implied the possibility of meaningful work outside of the home and independent living.

Women telegraphers realized that telegraphy as an occupation afforded opportunities for work that enabled them to use their minds as well as their hands and paid significantly more than traditional women's work. The opening of schools to teach telegraphy, together with the new middle-class attitudes about women working outside of the home, created opportunities for increasing numbers of women to enter the profession in the United States. In Europe, the nationalization of the telegraphs under the direction of the postal administration and the establishment of training programs for women created opportunities for women to enter the telegraph service as well.

Although male telegraphers periodically expressed concern about the consequences of giving women equal status in the telegraph office, the telegraph companies began to employ increasingly larger numbers of women telegraphers at lower rates of pay than the men. Women operators played an active role in the debate over their status that took place in the telegraph journals, and they demanded equal pay and privileges.

THE INDUSTRIALIZATION OF THE TELEGRAPH INDUSTRY

Not surprisingly, telegraphy was one of the earliest businesses to be industrialized. The industrialization of the telegraph business in the United States really took place in two phases, reflecting the fact that telegraphic work was performed in two separate workplace environments as part of two different industries—the communications industry, which employed the commercial operators, and the railroad industry, which employed the depot operators.

The work of the commercial operators began to be industrialized earlier, beginning in the 1870s, with the opening of large urban offices employing hundreds of operators. At first, women were given special status in this work environment and segregated into city departments or ladies' departments; later, they were integrated into the general workforce. Although the work was gendered to a degree through inequity in pay and status, the work was not fully stratified by gender until the introduction of the Teletype around 1915, when telegraphers were divided into the highly paid and predominantly male Morse operators and the predominantly female Teletype operators.

Industrialization in the commercial telegraph industry manifested itself through the elimination of benefits, longer working hours at reduced pay, and a more regimented work environment. As Shirley Tillotson points out in her article "'We may all soon be "first-class men,"'" the stress and pressure of the commercial office environment was a major differentiator between this environment and that of the single-person rural or depot office. The primary consequence of this industrialization was the creation of a new spirit of militancy among the telegraphers, who began to form labor unions. Their demands for higher pay and better working conditions led to a series of abortive and unsuccessful strikes in the late nineteenth and early twentieth centuries.[1]

Although women had been excluded from telegraphers' organizations in the preindustrial period, they were allowed to join the more activist unions and participated enthusiastically in the strikes of 1870, 1883, and 1907. The unions, realizing that nonunion women operators could create competition for jobs, gradually made women's issues a larger part of their agenda and included a demand for equal wages for men and women in 1883. Women's participation in the labor movement also led to alliances with other women's labor organizations and with the women's suffrage movement.

Industrialization of the telegrapher's work in the railroad industry did not occur until much later. Thus from the 1870s to the early twentieth century, two radically different work environments coexisted in the United States—the hectic pace of the commercial office, where messages were rushed back and forth by means of pneumatic tubes and clerks on roller skates, and the more leisurely but lower-paying environment of the depot office, where hours were irregular and the operator might have the opportunity to nap, or simply stare out the window, between train arrivals. Operators who had been blacklisted by Western Union as the result of a strike or who simply could not tolerate the regimentation of the city department frequently found refuge in the railroad depot.

In the railroad industry, industrialization of the work meant mechanization. The telephone and the Teletype eventually replaced the telegraph for the transmission of train orders, and the introduction of Centralized Train Control in the 1920s gradually eliminated the need for the Morse telegrapher. Orders no longer need to be "handed up" to trains but are sent via radio instead. While the job classification of "telegrapher" still exists in railroad work, the work no longer requires knowledge of Morse code; train routing today is heavily computerized and requires little human interaction.

The industrialization of the telegraph service in other parts of the world occurred somewhat later than in the United States and was driven by slightly different motivators. Toward the close of the nineteenth century, national legislatures in many parts of the world called for reductions in the rates charged by their respective telegraph administrations in an effort to promote the development of the communications sector and create a mass market. Telegraphic administrations responded to these rate reductions by implementing cost-cutting programs which generally included consolidation of services and increased employment of women at lower wages than their male counterparts. In France, this meant unification of the postal service and the telegraph service and the employment of increased numbers of women to operate post and telegraph offices. In Norway, the telephone and telegraph services were combined, and women who wished to enter the telegraph service first had to work in the lower-paying telephone service for a specified period of time. At the same time, men in the telegraph service were given additional opportunities for promotion to technical and administrative positions. This represented a deskilling of the positions available to women, who had previously been able to enter telegraphic service directly at a higher rate of pay; in the words of Gro Hagemann, "the ladder of employment was extended downwards for women and upwards for men." The eventual outcome was increased segregation by sex, as men moved into the highly paid technical and administrative positions and were replaced by women doing combined telephone and telegraph work at a lower rate of pay. Similar cost-cutting measures were undertaken in other postal and telegraphic administrations, with similar effects.[2]

Because wages and working conditions were fixed by government mandate, women operators in other parts of the world were less likely to organize into unions and demand changes in pay and working conditions than in the United States, where pay was generally determined on an individual basis. Requests for pay increases in European administrations generally required submitting a petition to parliament, followed by a slow and laborious legislative process. Commenting on the opposition that women operators faced when making demands for equality in pay and working conditions in Austria in 1919, Anna Rabenseifner noted, "The resistance that was [and is] offered by the bureaucrats in government service is naturally greater than that in private industry, because the reactionary spirit preserved itself best in the bureaus." Non-U.S. women operators also experienced arbitrary restrictions on their conditions of employment in some administrations, such as the requirement that they leave the service if they married.[3]

While women operators in the United States generally had more freedom to change jobs, to continue working after marriage, and to negotiate pay raises on an individual basis, their relationship with male operators was more competitive. Although women operators in the United States enthusiastically joined labor unions and participated in strikes, they might lose their jobs if the labor action failed. Thus the price that American telegraphers paid for greater freedom of action was less job security.

THE CLOSING OF THE TELEGRAPH AGE

By the 1920s, the age of the Morse telegraph operator was drawing to a close. The final phase of industrialization in the commercial telegraph office was the replacement of Morse operators with Teletype operators. For many telegraphers, the end came with the Depression that began in 1929; many of the Morse operators were simply laid off and replaced by Teletype operators. The Associated Press retired the last Morse operator in New York State in 1934. According to a press announcement on July 26, 1934: "The last of the 'brass pounders' has been displaced in the network of news wires of the Associated Press in New York State. To-day, from Niagara Falls to New York City, not a Morse operator is at work, automatic printers having replaced the last Morse circuit."[4]

The telephone industry also continued to make steady inroads into the telegraph business, as long-distance telephone service was perfected. Between 1926 and 1943, use of the telegraph declined 11 percent in the United States; during the same period, telephone usage increased by 39 percent.[5]

The decline was not caused simply by competition with the telephone. The telegraph industry had ceased to be technologically innovative.

Although *Wired Love*'s Nattie Rogers was aware of the possibilities of the facsimile process as early as the 1870s and Western Union possessed the technology to make it available to the general public by the early twentieth century, the telegraph company failed to understand the marketing potential of such service. Had it done so, it might have been able to anticipate the "fax boom" of the late 1980s. This was noted as early as 1945 by former Federal Communications Commission official Carrie Glasser, writing for the September 1945 issue of the *American Economic Review:* "The facsimile process of handling telegrams which does away with coding and decoding operations and thus reduces transmission time and reduces the possibilities for error, has long been known, but until recently steps were not taken by Western Union to place the process on a commercial basis."[6]

WOMEN TELEGRAPHERS AS TECHNICAL WORKERS

Although their occupational and social status were enigmatic to their contemporaries and contributed to the perception of women operators as "exceptional," telegraphers are readily recognizable as technical workers, or technicians, to modern eyes. With their desk jobs, obscure electrical apparatus, and complicated codes, telegraph operators seem to have more in common with modern-day technical workers than with most of their contemporaries.

Telegraphers were justifiably proud of their technical skills. Ma Kiley observed that "before I was allowed to take my first job I had to know the switchboard, how to ground wires, patch them, trace troubles, etc." Although not all women operators shared Ma Kiley's first-class knowledge of telegraph operation, all telegraphers had to know at a minimum how to send and receive Morse code and how to connect and adjust the instruments.[7]

The technological aspects of telegraphy force us to reconsider our assumptions that nineteenth-century electrical technology was a "man's world," in which women did not participate. The story of women in telegraphy shows that women not only participated in all aspects of telegraphic activity, from invention to management to the labor movement, but included telegraphy in the set of skills they passed to one another through ties based on shared profession, labor solidarity, and kinship.

Computer Programming and Telegraphy

The work of the telegrapher is related to that of modern computer programmers in surprising ways. Like the computer (and unlike the analog
telephone), the telegraph was a digital device that used dots and dashes in a
manner similar to the ones and zeroes of digital logic. One can think of the
telegraphers' instruments as "hardware" and Morse code as "software." The
telegrapher's work, like that of a modern computer programmer, consisted
of translating English-language instructions into machine-readable codes.
Morse code is, in fact, a direct ancestor of the American National Standard
Code for Information Interchange (ASCII) codes used by software programmers. The computer itself is the direct descendant of the telegraph; as
Carolyn Marvin observed in *When Old Technologies Were New:* "In a historical sense, the computer is no more than an instantaneous telegraph with a
prodigious memory, and all the communications inventions in between
have simply been elaborations on the telegraph's original work."[8] Like
women telegraphers, women computer programmers today constitute a significant minority in their profession; 1995 figures show 29.5 percent of all
computer programmers to be women.[9]

And despite tremendous gains by women in the work world, another
striking parallel exists between women computer programmers and women
telegraphers as well: the relative absence of their history in the profession.
Ruth Perry and Lisa Greber, writing on women's relationship to computers
in the autumn 1990 issue of *Signs,* noted that "research on the history of the
computer and its relationship to women still needs to be done because
much of the early history is missing. The currently available history under-
writes the standard story of the computer's masculine roots. The unwritten
history may tell a slightly different tale."[10]

Gender in Cyberspace—From the Telegraph to the Internet

Carolyn Marvin's metaphor of the computer as an instantaneous telegraph
has come to full fruition in the form of the Internet. Today, anyone with a
modem-equipped personal computer can use it as a personal telegraph to
send e-mail to anyone else in the world with similar equipment. Not surprisingly, widespread use of the Internet has led to a revisiting of some of

the same gender issues that telegraphers were familiar with a hundred years ago; reading about "Internet romances" in the popular press engenders a strong sense of déjà vu in anyone familiar with the story of telegraphy. We read about relationships that began with e-mail exchanges and "virtual weddings" conducted over the Internet, with the bride, groom, and minister all logging in at different locations. Stories of men and women masquerading as members of the opposite sex in Internet chat rooms are common; the debate over Internet "obscenity" and the alleged danger to public morals still rages.

The term "cyberspace" was coined to describe the psychic space a person inhabits when he or she is communicating electronically; the name is taken from the science fiction novel *Neuromancer*, written by William Gibson in 1984.[11] Cyberspace can be seen as a nongendered space in which all voices are disembodied; telegraphers, the original inhabitants of cyberspace, were the first to experience its gender-neutral qualities.

One of the earliest gender questions raised by telegraphers was whether there was a distinctively "female" style of sending; put in presentist language, was there a way of gendering voice in cyberspace and thereby recreating the traditional notion of separate spheres? The motivation was in essence ideological—which voices should be privileged and which devalued. Those who claimed that the distinctively "female" style of sending included clipping, a higher rate of errors, and a "tendency to jump at conclusions" were intent upon assigning a subordinate role to women in the communications industry.

Yet as the telegraphic romances hinted, the task of assigning worth in telegraphic cyberspace was not to be so simple. It was easy enough for men to masquerade as women, and women as men, over the line, by playing to the prejudices of the person at the other end. The fact that a sizable minority of first-class women operators made a career of press and market reporting, producing copy indistinguishable from that of their male peers, demonstrated the failure of any attempt to create a value system based on gender alone.

Thus the telegraph can be seen as subverting the predominant gender ideology of the age by creating a space (which we would call "cyberspace" today) that belonged to neither the predominantly masculine public sphere nor the predominantly feminine domestic sphere. Women telegraphers were able to explore, at least experimentally, a space in which their valuation was based on skill rather than appearance. Male operators as well had to adjust to a reality in which they were sometimes surprised to learn that the first-class "man" at the other end of the line was in fact a woman.

IN PURSUIT OF THE UNWRITTEN HISTORY

Few people today recall the role that the telegraph played in the past in providing ordinary people with rapid communications and in enabling trains to run safely and on time. Even fewer are aware of the role that women played in the telegraph industry.

Why was the story forgotten? Forgetting the story of women telegraph operators was a twentieth-century phenomenon, caused in part by methodological, if not ideological, bias on the part of business and labor historians. It is instructive to go through the telegraphic historical literature in chronological fashion and note first the appearance, then the gradual disappearance, and finally the reappearance of references to women. The nineteenth-century historians of the telegraph routinely discussed the presence of women in the industry. James D. Reid's 1879 history, *The Telegraph in America: Its Founders, Promoters, and Noted Men*, its title notwithstanding, provides a good account of the entry of women into the telegraph industry and provides biographies of several noted women telegraphers. William Plum's *Military Telegraph during the Civil War in the United States*, published three years later, also acknowledges the role of women operators in the Civil War.

Sometime in the early twentieth century, however, the role of women in the telegraph industry disappeared from the written history, and telegraphy began to be constructed as an archetypally male occupation. Both business history, as practiced in the early part of the twentieth century, and the John R. Commons school of labor history tended to focus on institutionalized power structures and to marginalize the role of women in the economic sector and in the labor market. Reflecting this bias, Robert L. Thompson's 1947 work, *Wiring a Continent: The History of the Telegraph Industry in the United States, 1832–1866*, considered to be the primary scholarly reference for the nineteenth-century telegraph industry, does not even mention that women worked as telegraphers. Similarly, in the field of labor history, Archibald McIsaac's 1933 study of railroad telegraphers, *The Order of Railroad Telegraphers: A Study in Trade Unionism and Collective Bargaining*, mentions the role of women only peripherally, as does Vidkunn Ulriksson's 1953 study of commercial telegraphers, *The Telegraphers: Their Craft and Their Union*.

It would be almost forty years before Edwin Gabler's 1988 book, *The American Telegrapher: A Social History, 1860–1900*, would appear, including a comprehensive study of women operators and their participation in the strike of 1883. Gabler's work reflected not only the emergence of social history as a historical genre but also the influence of the "new" labor history and its focus on workplace issues and the deskilling of work through industrialization.

The new interest in women's work, prompted in part by the women's movement of the 1970s, has also contributed to the literature on women in telegraphy. Melodie Andrews's essay "'What the Girls Can Do': The Debate over the Employment of Women in the Early American Telegraph Industry" discussed the entry of women into the U.S. telegraph industry in the pre–Civil War era and the debate over their status that appeared in the pages of the *Telegrapher* in the 1860s. Jacquelyn Dowd Hall's "O. Delight Smith's Progressive Era: Labor, Feminism, and Reform in the Urban South" explored the relationship between feminism, the progressive movement, and the labor movement exemplified in the life and work of Atlanta telegrapher Ola Delight Smith.

Historians in other countries as well have brought a feminist perspective to the study of women telegraphers, including Shirley Tillotson's study of Canadian women telegraphers, "'We may all soon be "first-class men"'": Gender and Skill in Canada's Early Twentieth Century Urban Telegraph Industry," and Gro Hagemann's study of the relationship between feminism and telegraphy in Norway, "Feminism and the Sexual Division of Labour: Female Labour in the Norwegian Telegraph Service Around the Turn of the Century."

Another reason for the lack of documentation of the work of women telegraphers relates to the technology itself. After the introduction of long-distance telephony around the turn of the century, the telegraph quickly faded from prominence. It became old-fashioned in an age that glorified progress, and thus its visible symbols were quickly tossed on the scrap heap—often quite literally. Telegraph equipment was scornfully demolished and thrown away in the early part of this century, and thus is now scarce and much sought after by collectors. Even the insulators from the telegraph era are genuine rareties, selling for as much as several hundred dollars apiece.

Corporate Recordkeeping

As is typical in other fields of women's work, telegraph companies did not keep accurate employment records because women were not considered to be serious workers. Corporate records on the subject are scarce. Most large companies did not begin to keep centralized personnel records of their employees until well into the twentieth century. This was true of the Western Union Telegraph Company and many railroads as well. Western Union knew at any time how many employees it had, but to find out the names and pay rates of individual employees, one would have to go to each individual office and look at the office ledger book. Few of these ledger books survive today. Of the few records that were generated initially, even fewer survive today. Many of the railroads destroyed personnel records during the close-downs that occurred during the cutbacks of the 1960s–1980s.

The railroads may also have been reluctant to document actual working hours for women workers after the passage of the eight-hour laws in the early 1900s. Since they would have been liable for fines if it could be proven that women operators were required to work more than eight hours a day, some railroads may have simply avoided documenting the presence of women on the payroll.

The Role of the Telegraphers' Journals

Although their story is largely missing from the history books and the corporate records, information about women telegraphers can be found in the telegraphers' and railroaders' journals. Trade journals from the nineteenth century such as the *Telegrapher, Telegraph Age*, and the *Operator* provide valuable information on the work, lives, and opinions of women telegraphers; union journals, like the *Commercial Telegraphers' Journal* of the CTUA and the *Railroad Telegrapher* also help to fill in the missing history. *Railroad Magazine* and its predecessors, *Railroad, Railroad Stories*, and *Railroad Man's Magazine*, frequently published the life stories and recollections of male and female railroad telegraphers. Women operators were frequent and vocal contributors to these journals; they used the telegraphers' journals to debate the status of women in the telegraph industry.

Researching the Work and Lives of Telegraph Operators

As Ava Baron notes in her essay "Gender and Labor History: Learning from the Past, Looking to the Future," researching the lives of nineteenth- and early twentieth-century women workers, including telegraphers, requires a different approach from that used to study their male contemporaries. Nontraditional sources and techniques borrowed from genealogy are helpful in this work. Women operators tended to leave their traces in birth, death, and marriage records, while men are more commonly recorded in land transactions, court records, and military service records. In a "traditional" sense, birth and marriage records are used by genealogists, while historians rely on legislative, legal, and military records. Of course this has changed with the introduction of social history and women's history. And the following paragraph adds a few more examples.[12]

The stories of women telegraph operators can be found in railroad and telegraphic journals, in census records, and in their obituaries. Railroad Retirement Board records are also a valuable source of information on telegraphers who retired after its establishment in 1935. Local newspapers and historical societies often can provide information on the lives of telegraphers in their communities. Perhaps the most important sources of information, however, are the oral histories of the operators themselves. The recollections of retired telegraph operators are rich sources of facts and observations.

THE IMPORTANCE OF REDISCOVERING HISTORY

The story of women telegraph operators provides insight into the role that women played in the growth and expansion of technology in the nineteenth and early twentieth centuries. While the current historical record incorrectly suggests that women did not encounter electricity until it appeared in the home in the form of appliances, the story of women in telegraphy indicates that women played an active role in the development of electrically based communications technology. Women as telegraph operators broke new ground by creating new employment opportunities and helped to shape the technology that in turn created the modern "global village." Despite prejudices against their employment, female telegraphers provided their communities with instantaneous communications and enabled the trains to run safely and on time. Thus rediscovering the history of women in telegraphy serves a dual purpose: not only does it illuminate a little-understood area of nineteenth-century women's work, it also gives us a deep historical perspective on the role of women in technology and how women in the past have sought to gain control over their professional lives and recognition in their fields. We can better understand the future by reclaiming the past.

Notes

CHAPTER 1—WOMEN IN THE TELEGRAPH INDUSTRY

1. Frances E. Willard, *Occupations for Women* (New York, 1897), 132, quoted in Edwin Gabler, *The American Telegrapher: A Social History, 1860–1900* (New Brunswick, N.J.: Rutgers University Press, 1988), 108.

2. "Women as Telegraph Operators," *Electrical World*, June 26, 1886, 296.

3. Shirley Tillotson, "'We may all soon be "first-class men"': Gender and Skill in Canada's Early Twentieth Century Urban Telegraph Industry," *Labour/Le Travail* 27 (spring 1991): 98.

4. Bernice Selden, *The Mill Girls* (New York: Atheneum, 1983), 174. See also Helena Wright, "Sarah G. Bagley: A Biographical Note," *Labor History* 20 (summer 1979): 398–413.

5. J. J. Speed to Ezra Cornell, July 13, 1849, Box 10, Folder 3, Ezra Cornell Papers, Division of Rare and Manuscript Collections, Cornell University, Ithaca, New York.

6. Phoebe Wood to Ezra Cornell, September 23, 1849, Box 10, Folder 8, ibid.

7. Phoebe Wood to Ezra Cornell, November 24, 1849, Box 10, Folder 11, ibid.

8. "The First Woman Operator," *Telegraph and Telephone Age*, October 1, 1910, 659–60; James D. Reid, *The Telegraph in America: Its Founders, Promoters, and Noted Men* (New York: Derby Brothers, 1879), 170–71; *West Chester (Pa.) Daily Local News*, December 22, 1904.

9. *The American Railway: Its Construction, Management, and Appliances*, intro. Thomas M. Cooley (1897; reprint, New York: Arno Press, 1976); see also Ian R. Bartky, "Running on Time," *Railroad History* 159 (autumn 1988), 18–38; and Robert L. Thompson, *Wiring a Continent: The History of the Telegraph Industry in the United States, 1832–1866* (Princeton: Princeton University Press, 1947), 206–9.

10. "Aged Lady's Fall Causes Death," *Lewistown (Pa.) Sentinel*, March 24, 1922; "The Oldest Lady Telegrapher," *Telegraph Age*, September 16, 1897; ibid., August 16, 1907, 445.

11. "The First Woman Operator"; *Telegraph and Telephone Age*, October 1, 1910, 659–60; March 1, 1907, 142; "Woman Ran First Lynn-Boston Wire," *Boston Herald*, December 8, 1907.

12. Virginia Penny, *How Women Can Make Money* (Springfield, Mass.: Fisk, 1870), 100–101.

13. For the employment of women in the Canadian telegraph industry, see Shirley Tillotson, "The Operators along the Coast: A Case Study of the Link between Gender, Skilled Labour and Social Power, 1900–1930," *Acadiensis* 20 (1990): 72–88; Tillotson, "We may all soon be 'first-class men.'"

14. The employment of women in the telegraph industry in England and Europe is described in Jeffrey Kieve, *The Electric Telegraph: A Social and Economic History* (Newton Abbot, Devon: David & Charles, 1973), 39, 85; Charles Garland, "Women as Telegraphists," *Economic Journal*, June 1901, 252, 255; and Jeanne Bouvier, *Histoire des dames employées dans les postes, télégraphes et téléphones de 1714 à 1929* (Paris: Presses Universitaires de France, 1930), 128–30.

15. For the story of undersea cables, see Jorma Ahvenainen, "The Far Eastern Telegraphs: The History of Telegraphic Communications between the Far East, Europe and America before the First World War," *Suomalaisen Tiedeakatemian Toimituksia*, Sarja-Ser. B, Nide-Tom 216 (1981): 32–35, 39, 41; and Erik Baark, *Lightning Wires: The Telegraph and China's Technological Modernization, 1860–1890* (Westport, Conn.: Greenwood Press, 1997). For the building of the overland lines to Asia by the Siemens brothers, see A. Karbelashvily, "Europe-India Telegraph 'Bridge' via the Caucasus," *Telecommunications Journal* 56 (1989): 719–23. For the employment of women as telegraphers in India, see Krishnalal Shridharani, *Story of the Indian Telegraphs: A Century of Progress* (New Delhi: Government of India Press, 1953), 65. See also Garland, "Women as Telegraphists," 252.

16. Griffith Taylor, *Australia: A Study of Warm Environments and Their Effect on British Settlement* (New York: Dutton, 1932), 373; Garland, "Women as Telegraphists," 252.

17. Ahvenainen, "Far Eastern Telegraphs," 17; Gouvernement Général de L'Afrique Occidentale Française, *Les postes et télégraphes en Afrique Occidentale* (Corbeil, France: Ed. Crete, Imprimerie Typographique, 1907), 6–28; *Encyclopedia of Southern Africa*, ed. Eric Rosenthal (London: Frederick Warne, 1973), 574; Garland, "Women as Telegraphists," 252.

18. "Mrs. Abbie Vaughan 'Mother of Code Telegraphy' Dies at Home Here," *Long Beach (Calif.) Press*, August 19, 1924; Thomas C. Jepsen, *Ma Kiley: The Life of a Railroad Telegrapher* (El Paso: Texas Western Press, 1997).

19. John J. Johnson, "Pioneer Telegraphy in Chile, 1852–1876," *Stanford University Publications University Series, History, Economics, and Political Science* 6, no. 1 (1948): 88; Garland, "Women as Telegraphists," 252.

20. Roger Fison, "Is Morse Telegraphy Doomed to Extinction?" *Railroad Man's Magazine* 3 (May 1917): 60–77; Elizabeth Faulkner Baker, *Technology and Women's Work* (New York: Columbia University Press, 1964), 244.

CHAPTER 2—DAILY LIFE IN THE TELEGRAPH OFFICE

1. "The Telegraph," *Harper's Magazine*, August 1873, 347.

2. Ella Cheever Thayer, *Wired Love* (New York: W. J. Johnston, 1880), 12; Josie Schofield, "Wooing by Wire," *Telegrapher*, November 20, 1875, 277.

3. Descriptions of the various types of telegrams are given in Ralph Edward Berry, *The Work of Juniors in the Telegraph Service* (Berkeley: University of California Division of Vocational Education, 1922), 134–35.

4. "The Story of Telegraphy," comp. Kate B. Carter, from *Our Pioneer Heritage* (Salt Lake City: Daughters of Utah Pioneers, 1961), 4:549–50.

5. "Miss Medora Olive Newell, Postal Manager in Chicago," *Telegraph Age*, June 1, 1909, 396; ibid., August 1, 1905, 300.

6. Jepsen, *Ma Kiley*, 97.

7. Sue R. Morehead, "Woman Op," *Railroad Magazine*, January 1944, 89–90.

8. Roger Reinke, "Telegraph Equipment Classification," *Antique Wireless Association Old Timer's Bulletin* 32 (May 1991): 35–37.

9. Jepsen, *Ma Kiley*, 47.

10. Job descriptions are taken from Berry, *Work of Juniors in the Telegraph Service*, 25–56.

11. Richard O'Connor, *Johnstown: The Day the Dam Broke* (New York: J. B. Lippincott, 1957), 60; *Johnstown (Pa.) Daily Tribune*, February 27, 1940.

12. Gabler, *American Telegrapher*, 122; *Journal of the Telegraph*, May 1, 1869, 134; *Telegrapher*, April 3, 1875, 80.

13. "New Western Union Chief Operator at Springfield, Mass.," *Telegraph Age*, May 1, 1905, 180.

14. See, for example, *Telegrapher*, November 28, 1864, 20.

15. Carter, "Story of Telegraphy," 561.

16. *West Chester (Pa.) Daily Local News*, December 22, 1904; Carter, "Story of Telegraphy," 553–54.

17. Gabler, *American Telegrapher*, 113, 140; Vidkunn Ulriksson, *The Telegraphers: Their Craft and Their Unions* (Washington, D.C.: Public Affairs Press, 1953), 101; *New York Times*, July 21, 1944.

18. Quoted in Gabler, *American Telegrapher*, 55.

19. Barnet Phillips, "The Thorsdale Telegraphs," *Atlantic Monthly*, October 1876, 401.

20. "Lady Operators," *Telegrapher*, February 27, 1865, 58.

21. Lee Holcombe, *Victorian Ladies at Work: Middle-Class Working Women in England and Wales, 1850–1914* (Hamden, Conn.: Archon Books, 1973), 166.

22. *Telegrapher*, November 18, 1871, 99.

23. Ibid., March 6, 1875, 59.

24. Ibid., February 13, 1875, 38.

25. Ambrose Gonzales to Willie, October 15, 1881, quoted in Thomas C. Jepsen, "Two 'Lightning Slingers' from South Carolina," *South Carolina Historical Magazine* 94 (October 1993): 276; Charles Buckingham, "The Telegraph of To-Day," *Scribner's Magazine*, July 1889, 8.

26. *Operator*, August 15, 1882, 343.

27. *Chicago Tribune*, August 18, 1907.

28. Gabler, *American Telegrapher*, 111; Ulriksson, *Telegraphers*, 99.

29. William R. Plum, *The Military Telegraph during the Civil War in the United States*, 2 vols. (Chicago: Jansen, McClurg, 1882), 1:345–46; *Telegrapher*, February 26, 1876, 51.

30. Garland, "Women as Telegraphists," 258–59.

31. *Operator*, April 15, 1882, 155.

32. Some railroads evidently considered forbidding the employment of women rather than deal with the logistics of fitting an eight-hour workday into their train schedules.

According to the *Railroad Telegrapher* of November 1910 (page 1705), "There is a rumor that all the railroads may follow the example of the B. & O. and bar women employees." But the railroads continued to employ increasing numbers of women employees until the 1920s, when automation began to reduce the ranks of telegraphers.

33. The Teletype and its printer are the direct ancestors of modern computer keyboards and printers; designers of early computers simply appropriated Teletype devices to serve as input/output devices. Thus, in a sense, computers appropriated the language of the telegraph to communicate with humans.

34. Fison, "Is Morse Telegraphy Doomed to Extinction?" 71.

35. Ibid., 67.

36. Lucile Ross, "Railroads," in *Our Life, 1882–1982, Akron, Iowa* (Akron, Iowa: *Akron Register-Tribune* and *Le Mars Daily Sentinel* Job Printing, 1982), 30.

37. *Telegrapher*, July 15, 1876, 173; July 22, 1876, 180.

38. N. G. Gonzales to Emily Elliott, quoted in Jepsen, "Two 'Lightning Slingers' from South Carolina," 272; *Telegraph Age*, September 1, 1897, 316.

39. *Telegrapher*, September 15, 1869, 272; September 23, 1876, 234; Jepsen, *Ma Kiley*, 58.

40. Martha L. Rayne, *What Can a Woman Do? or, Her Position in the Business and Literary World* (Petersburgh, N.Y.: Eagle, 1893), 140–41; see also Elizabeth Beardsley Butler, *Women and the Trades: Pittsburgh, 1907–8* (Pittsburgh: University of Pittsburgh Press, 1984), 293.

CHAPTER 3—SOCIETY AND THE TELEGRAPH OPERATOR

1. Thayer, *Wired Love*, 25; Minnie Swan Mitchell, "Lingo of Telegraph Operators," *American Speech*, April 1937, 155.

2. Gabler, *American Telegrapher*, 173.

3. "Female School of Telegraphy," *Journal of the Telegraph*, November 1, 1869, 271.

4. Alexander H. Bullock, "The Centennial Situation of Woman," *Address of Hon. Alexander H. Bullock at the Commencement Anniversary of Mount Holyoke Seminary, Massachusetts, June 22, 1876* (Worcester, Mass.: C. Hamilton, 1876), 6.

5. Edwin Gabler discusses the class status of telegraphers in *The American Telegrapher*, 85–91 and 123. Stephen Meyer discussed the concept of "rough masculine culture" in the workplace in "Work, Play, and Power: Masculine Culture on the Automotive Shop Floor," a paper presented at "Boys and Their Toys? Masculinity, Technology, and Work," a conference held at the Center for the History of Business, Technology, and Society, Hagley Museum and Library, in Wilmington, Delaware, on October 3, 1997.

6. For Sarah Bagley's class origins, see Wright, "Sarah G. Bagley," 398–413, and Hannah Josephson, *The Golden Threads: New England's Mill Girls and Magnates* (New York: Duell, Sloan and Pearce, 1949), 250–74; Mary Stillwell's background is described in Robert Conot, *A Streak of Luck* (New York: Seaview Books, 1979), 46–48, 219.

7. Kieve, *Electric Telegraph*, 85; Garland, "Women as Telegraphists," 251.

8. Gro Hagemann, "Feminism and the Sexual Division of Labour: Female Labour in the Norwegian Telegraph Service around the Turn of the Century," *Scandinavian Journal of History* 10 (1985): 151.

9. Thayer, *Wired Love*, 54; "The Lady Operators," *New York World*, July 27, 1883, 2.

10. Gabler, *American Telegrapher*, 119.

11. *Telegrapher Supplement*, November 6, 1865, 13; Jepsen, "Two 'Lightning Slingers' from South Carolina," 264–82; *Telegrapher*, February 1, 1873, 32.

12. *1890 Census*, vol. 2, pt. 2, table 82, "Total Persons 10 Years of Age and Over in the United States Engaged in Each Specified Occupation, Classified by Sex, General Nativity, and Color," 356.

13. Ibid., table 111, "White Persons 10 Years of Age and Over in the United States Engaged in Each Specified Occupation, Classified by Sex and Birthplace of Mothers," 504.

14. Margery Davies, *Woman's Place Is at the Typewriter* (Philadelphia: Temple University Press, 1982), 57; see also table 2, appendix.

15. *Journal of theTelegraph*, January 15, 1869, 42.

16. "The Cooper Union—Telegraph School for Women: Rules and Regulations for Its Government," *Journal of theTelegraph*, February 15, 1869, 70.

17. Ibid., November 1, 1869, 271.

18. Ibid., October 16, 1871, 267.

19. Gabler, *American Telegrapher*, 113–14, 132–33.

20. *Annals of Cleveland*, vol. 48, 1865 (Cleveland: Works Progress Administration in Ohio, 1938), 47; *Telegrapher*, March 27, 1865, 70.

21. Ibid., December 20, 1873, 309.

22. *Annals of Cleveland*, vol. 57, 1874 (Cleveland: Works Progress Administration in Ohio, 1937), 742.

23. *Operator*, April 15, 1882, 156.

24. Jepsen, *Ma Kiley*, 56.

25. *Lewistown (Pa.) Sentinel*, March 24, 1922; Jepsen, *Ma Kiley*, 55–56.

26. Carter, "Story of Telegraphy," 566.

27. Correspondence, Lynne Belluscio to author Le Roy Historical Society, January 14, 1993.

28. Fison, "Is Morse Telegraphy Doomed to Extinction?" 66; interview with author, July 3, 1997.

29. Kieve, *Electric Telegraph*, 85; "Young Women at the London Telegraph Office," *New York Times*, June 17, 1877.

30. Bouvier, *Histoire des dames employées dans les postes, télégraphes et téléphones*, 130–32.

31. Garland, "Women as Telegraphists," 254–55.

32. Johnson, "Pioneer Telegraphy in Chile," 88–89.

33. Rayne, *What Can a Woman Do?* 139.

34. Kieve, *Electric Telegraph*, 87.

35. Garland, "Women as Telegraphists," 251.

36. Data for tables 1 and 2 are taken from the occupation tables of the U.S. Census for 1870–1960. The Northeast/Mid-Atlantic region includes Massachusetts, Connecticut, Maine, Ohio, Maryland, New Hampshire, Vermont, the District of Columbia, New Jersey, New York, Pennsylvania, Rhode Island, Delaware, and West Virginia; the South encompasses Alabama, Florida, Georgia, Kentucky, Louisiana, Mississippi, North Carolina, South Carolina, Tennessee, and Virginia; the Midwest/Prairie region includes Indiana, Illinois, Minnesota, Iowa, Missouri, Wisconsin, and Michigan; the Far West/Plains region includes North Dakota, South Dakota, Nebraska, Kansas,

Oklahoma, Arkansas, Texas, Idaho, Utah, Nevada, California, Colorado, Wyoming, Montana, Oregon, Washington, Arizona, New Mexico, Alaska, and Hawaii.

37. Margo Anderson, "The History of Women and the History of Statistics," *Journal of Women's History* 4 (spring 1992): 22. Census takers were well aware of the discrepancies, as noted in volume 1 of the 1870 census (p. 663): "A single sample will suffice. The Tables of Occupations characterize but a little over two thousand persons as connected with the woolen and worsted manufactures, while the Tables of Manufactures show that considerably more than forty thousand persons were engaged, upon the average, in these branches of manufacturing industry."

38. *Chicago Tribune*, August 15, 1883; *New York Herald*, July 28, 1883.

39. Archibald M. McIsaac, *The Order of Railroad Telegraphers: A Study in Trade Unionism and Collective Bargaining* (Princeton: Princeton University Press, 1933), 3.

40. Leo Wolman, *The Growth of American Trade Unions, 1880–1923* (New York: National Bureau of Economic Research, 1929), 99.

41. "Lady Clerks," *Anson Times* (Wadesboro, N.C.), April 17, 1884; for a discussion of the employment of women in the South in the post–Civil War era, see Edward L. Ayers, *The Promise of the New South* (New York: Oxford University Press, 1992), 77–79.

42. *Anson Times*, July 6, 1882.

43. Jacquelyn Dowd Hall, "O. Delight Smith's Progressive Era: Labor, Feminism, and Reform in the Urban South," in *Visible Women: New Essays on American Activism*, ed. Nancy A. Hewitt and Suzanne Lebsock (Champaign: University of Illinois Press, 1993) 169.

44. "Women as Telegraph Operators," *Electrical World* 8 (June 26, 1886): 296.

45. Carter, "Story of Telegraphy," 534, 541, 546. The Deseret Telegraph was incorporated as a company in 1867; it remained in existence until 1900, when it was acquired by Western Union.

46. Tillotson, "'We may all soon be "first-class men,"'" 98–99.

47. Bouvier, *Histoire des dames employées dans les postes, télégraphes et téléphones*, 128.

48. Ibid., 132.

49. Anna Rabenseifner, "Die Frau im oeffentlichen Dienst," *Handbuch der Frauenarbeit in Oesterreich* (Vienna: Carl Ueberreuter, 1930), 228–29.

50. Thayer, *Wired Love*, 28; Gabler, *American Telegrapher*, 108–9; *Lewistown (Pa.) Sentinel*, March 24, 1922; *Long Beach (Calif.) Press*, August 19, 1924.

51. "Miss Medora Olive Newell, Postal Manager in Chicago," *Telegraph Age*, June 1, 1909, 396; correspondence, Belluscio to author, Le Roy Historical Society, January 14, 1993; *New York Times*, July 21, 1944, 19.

52. Garland, "Women as Telegraphists," 259.

53. Louise R. Moreau, "The Feminine Touch in Telecommunications," *AWA Review* 4 (1989): 73; *Telegraph and Telephone Age*, October 1, 1910, 659–60; Penny, *How Women Can Make Money*, 101.

54. Pay rates for male and female operators are discussed in Gabler, *American Telegrapher*, 71, 94–95, 109, 112; Rayne, *What Can a Woman Do?* 143; Butler, *Women and the Trades*, 294; Jepsen, *Ma Kiley*, 82; and "The Convention Attended by Delegates from Local No. 16, C. T. U. A.," *Commercial Telegraphers' Journal: The Official Organ of the Commercial Telegraphers' Union of America* 5 (August 1907): 848.

55. *Harrisburg Ledger*, Series 3, Subseries 1, Box 11, Folder 2, Western Union Col-

lection, National Museum of American History, Smithsonian Institution, Washington, D.C.

56. Gabler, *American Telegrapher*, 124–25; William F. G. Shanks, "Women's Work and Wages," *Harper's Magazine*, September 1868, 546–53; Penny, *How Women Can Make Money*, 38, 181, 426; Rayne, *What Can a Woman Do?* 18.

57. John Brooks, *Telephone: The First Hundred Years* (New York: Harper & Row, 1976), 134, 66; Paul H. Douglas, *Real Wages in the United States, 1890–1926* (Boston: Houghton Mifflin, 1930), 331; *Commercial Telegraphers' Journal* 43 (January 1945): 4.

58. Gabler, *American Telegrapher*, 135; Baker, *Technology and Women's Work*, 68; Butler, *Women and the Trades*, 294; Rayne, *What Can a Woman Do?* 137.

59. Rayne, *What Can a Woman Do?* 139–41; *Report of the Michigan State Commission of Inquiry into Wages and the Conditions of Labor for Women and the Advisibility of Establishing a Minimum Wage* (Lansing, Mich.: Wynkoop Hallenbeck Crawford, 1915), 136.

60. *Commercial Telegraphers' Journal* 35 (December 1937): 142.

61. Tillotson, "Operators along the Coast," 79; Tillotson, "'We may all soon be "first-class men,"'" 107.

62. Kieve, *Electric Telegraph*, 87; Garland, "Women as Telegraphists," 253. I am using an 1866 exchange rate of 1 pound = $4.86.

63. Bouvier, *Histoire des dames employées dans les postes, télégraphes et téléphones*, 132. I am using an 1869 exchange rate of 7 francs = $1.00.

64. Shridharani, *Story of the Indian Telegraphs*, 65.

65. *Telegrapher*, April 25, 1865, 83; February 26, 1876, 49; July 29, 1876, 181.

66. *Telegraph Age*, March 16, 1893, 107; April 1, 1893, 119–22; April 16, 1893, 188; August 16, 1894, 325.

67. Minerva C. Smith, "Does It Pay a Girl to Learn Telegraphy?" *Commercial Telegraphers' Journal* 5 (November 1907): 1191–92.

68. *Vinton (Iowa) Eagle*, December 2, 1874; reprinted in *Telegrapher*, January 2, 1875, 1; ibid., September 4, 1875, July 1, 1876.

69. *Telegrapher*, March 6, 1875, 57; November 6, 1875, 267; Rayne, *What Can a Woman Do?* 139.

70. "Miss Medora Olive Newell, Postal Manager in Chicago," *Telegraph Age*, June 1, 1909, 396.

71. "Mrs. M. E. Randolph," *Telegraph Age*, March 9, 1893, 85; correspondence, Belluscio to author, Le Roy Historical Society, January 14, 1993.

72. "The Women's League of the Western Union Telegraph Company: Report for the Year 1918–1919," Box 47, Folder 3, Western Union Collection.

73. Hall, "O. Delight Smith's Progressive Era," 173–74; *Railroad Telegrapher*, May 1909, 700; *Telegraph Age*, June 1, 1909, 396.

74. Wright, "Sarah G. Bagley," 398–413; *New York Times*, July 21, 1944, 19.

75. *New York Times*, April 11, 1890.

76. "Sketches of Some of the Champions of the Philadelphia Tournament," *Telegraph Age*, January 1, 1904, 14.

77. Carole Turbin, "Reconceptualizing Family, Work, and Labor Organizing: Working Women in Troy, 1860–1890," in *Hidden Aspects of Women's Work*, ed. Christine Bose, Roslyn Feldberg, Natalie Sokoloff (New York: Praeger, 1987), 181; *Telegrapher*, February 1, 1866, 42.

78. "The Telegraphic Education of Women," *Journal of the Telegraph*, December 15, 1870, 22.

79. Alice Kessler-Harris, *A Woman's Wage: Historical Meanings and Social Consequences* (Lexington: University Press of Kentucky, 1990), 42; Carter, "Story of Telegraphy," 549–50; "The Lady Operators," *New York World*, July 27, 1883, 2.

80. "The Western Union Female Telegraphers Done in Rhyme by One of Themselves," *Telegrapher*, March 6, 1875, 57; *New York World*, July 24, 1883; *Telegrapher*, March 23, 1872, 242.

81. *Long Beach (Calif.) Press*, August 19, 1924.

82. *Johnstown (Pa.) Daily Tribune*, February 27, 1940; Jepsen, *Ma Kiley*, 53–55; *Washington Post*, August 10, 1907.

83. Carter, "Story of Telegraphy," 566, 562.

84. "The Lady Operators," *New York World*, July 27, 1883, 2.

85. Carter, "Story of Telegraphy," 554–55.

86. Shirley Burman, curator, "Women and the American Railroad: 135 Years of Womens' Association with the Railroad"; correspondence, William Strobridge, Research Associate, Wells Fargo Bank, December 1, 1992.

87. *Long Beach (Calif.) Press*, August 19, 1924.

88. Jepsen, *Ma Kiley*, 59, 66–67.

CHAPTER 4—WOMEN'S ISSUES IN THE TELEGRAPH OFFICE

1. Philip Foner, *Women and the American Labor Movement: From Colonial Times to the Eve of World War I* (New York: Free Press, 1979), 109.

2. *Lewistown (Pa.) Sentinel*, March 24, 1922; *Long Beach (Calif.) Press*, August 19, 1924.

3. *Telegraph Age*, March 9, 1893, 85; April 16, 1909, 301.

4. *Johnstown (Pa.) Daily Tribune*, February 27, 1940; David McCullough, *The Johnstown Flood* (New York: Simon and Schuster, 1968), 97.

5. Moreau, "Feminine Touch in Telecommunications," 73. For a good description of the Confederate use of the telegraph, see J. Cutler Andrews, "The Southern Telegraph Company, 1861–1865: A Chapter in the History of Wartime Communication," *Journal of Southern History* 30 (1964): 319–44.

6. Plum, *Military Telegraph during the Civil War in the United States*, 2:218–19; David Homer Bates, *Lincoln in the Telegraph Office* (New York: Century, 1907), 14–37; *Congressional Record*, vol. 29, pt. 2 (January 28, 1897), 1243; *Telegraph Age*, June 1, 1909, 380; July 1, 1909, 498.

7. Penny, *How Women Can Make Money*, 101.

8. Foner, *Women and the American Labor Movement*, 110–14.

9. Thompson, *Wiring a Continent*, 389–90, 392; Gabler, *American Telegrapher*, 150.

10. *Telegrapher*, October 31, 1864, 16. Melodie Andrews discusses the entry of women into the telegraph industry during the Civil War era and the subsequent debate in the *Telegrapher* in "'What the Girls Can Do': The Debate over the Employment of Women in the Early American Telegraph Industry," *Essays in Economic and Business History* 8 (1990): 109–20.

11. *Telegrapher*, November 28, 1864, 20; see also Cindy Sondik Aron, *Ladies and Gen-*

tlemen of the Civil Service: Middle-Class Workers in Victorian America (New York: Oxford University Press, 1987), 70–78.

12. *Telegrapher*, December 26, 1864, 32.

13. Ibid., January 30, 1865, 49; *Rules and Instructions for the Information and Guidance of the Employees of the Western Union Telegraph Company* (New York: Russell Brothers, 1870), 21; *Telegrapher*, January 30, 1865, 48.

14. *Telegrapher*, February 27, 1865, 61, 62, 58.

15. *Telegrapher Supplement*, November 6, 1865, 12–13.

16. *Telegrapher*, November 1, 1865, 184; *New York Times*, November 26, 1865.

17. *Telegrapher*, January 15, 1866, 33.

18. Ibid., February 1, 1866, 42.

19. Ibid., March 1, 1866, 68.

20. Ibid., September 21, 1867, 32; June 13, 1868, quoted in ibid., August 3, 1872, 396.

21. Ulriksson, *Telegraphers*, 18–20.

22. Andrews, "'What the Girls Can Do,'" 109–20.

23. Bouvier, *Histoire des dames employées dans les postes, télégraphes et téléphones*, 127–30.

24. *Telegrapher*, January 9, 1875, 9.

25. "The Telegraph and the Business Depression—Reduction of Salaries and Over Supply of Operators," *Telegrapher*, November 29, 1873, 292. Surprisingly, while lamenting the oversupply of operators and competition for jobs, this editorial makes no mention of competition between men and women for employment.

26. Ibid., January 9, 1875, 9.

27. Ibid., January 23, 1875, 20.

28. Ibid., March 27, 1875, 74.

29. Stephen Meyer, "Work, Play, and Power: Masculine Culture on the Automotive Shop Floor."

30. *Telegrapher*, January 30, 1865, 49; Jepsen, *Ma Kiley*, 91.

31. *Rules and Instructions for the Information and Guidance of the Employees of the Western Union Telegraph Company*, 21.

32. "A Fallen Angell," *Telegrapher*, November 6, 1875, 266; ibid., December 4, 1875, 290.

33. Ibid., November 6, 1875, 266.

34. Ibid., December 4, 1875, 290.

35. *Chicago Tribune*, August 6, 1883.

36. Rayne, *What Can a Woman Do?* 138; *Telegrapher*, February 27, 1865, 58.

37. Garland, "Women as Telegraphists," 257.

38. H. M. Cammon, "Women's Work and Wages," *Harper's Magazine*, April 1869, 665–70.

39. *Telegrapher*, August 3, 1872, 396.

40. *Operator*, May 15, 1882, 198.

41. *New York World*, July 27, 1883, 2.

42. Jepsen, *Ma Kiley*, 66.

43. Hagemann, "Feminism and the Sexual Division of Labour," 143–47.

44. *Telegrapher*, February 1, 1866, 42.

45. McCullough, *Johnstown Flood*, 93–105; O'Connor, *Johnstown, The Day the Dam*

Broke, 60–61; correspondence, Robin Rummel to author, Johnstown Area Heritage Association, 1991.

46. *Telegrapher*, July 22, 1871, 382. Melodie Andrews mentions this example of female entrepreneurship in the telegraph industry in "'What the Girls Can Do.'"

47. Paul Israel, *From Machine Shop to Industrial Laboratory: Telegraphers and the Changing Context of American Invention, 1830–1920* (Baltimore: Johns Hopkins University Press, 1992), 82.

48. Aron, *Ladies and Gentlemen of the Civil Service*, 144.

49. *Telegrapher*, June 17, 1871, 337–40.

50. Ibid., February 19, 1870, 208.

51. Ibid., March 23, 1872, 242.

52. Ibid., March 6, 1875, 59.

53. "Woman Is First of Sex to Win Executive Post in Western Union," *New York Herald Tribune*, July 3, 1925, 3.

54. Gabler, *American Telegrapher*, 131–32.

55. Hall, "O. Delight Smith's Progressive Era," 171.

56. *Telegrapher*, March 25, 1876, 75.

57. Jepsen, *Ma Kiley*, 75.

58. Margaret Dreier Robins to Mary Dreier, September 12, 1907, *The Papers of the Women's Trade Union League and Its Principal Leaders* (microfilm), ed. Edward T. James (Woodbridge, Conn.: Research Publications, 1981), reel 20.

59. Mara Keire, "The Vice Trust: A Reinterpretation of the White Slavery Scare in the United States, 1907–1917," Research Seminar Paper 48, Center for the History of Business, Technology, and Society, Hagley Museum and Library, November 6, 1997.

60. "The Dangers of Wired Love," *Electrical World*, February 13, 1886, 68–69. Carolyn Marvin discusses Maggie McCutcheon in *When Old Technologies Were New* (New York: Oxford University Press, 1989), 74.

CHAPTER 5—WOMEN TELEGRAPHERS IN LITERATURE AND CINEMA

1. Amy Gilman discusses the treatment of nineteenth-century working women in fiction in "'Cogs to the Wheels': The Ideology of Women's Work in Mid-19th-Century Fiction," in *Hidden Aspects of Women's Work*, ed. Christine Bose, Roslyn Feldberg, and Natalie Sokoloff (New York: Praeger, 1987), 116–34.

2. Mitchell, "Lingo of Telegraph Operators," 154–55.

3. Justin McCarthy, "Along the Wires," *Harper's New Monthly Magazine*, February 1870, 416–20.

4. The supposed threat to femininity and moral superiority posed by the workplace is discussed in Davies, *Woman's Place Is at the Typewriter*, 80–81.

5. Gilman, "'Cogs to the Wheels,'" 119.

6. Phillips, "Thorsdale Telegraphs," 400–417.

7. Schofield, "Wooing by Wire," *Telegrapher*, November 20, 1875, 277–78, November 27, 1875, 283–84. The story also appeared in *Lightning Flashes and Electric Dashes: A Volume of Choice Telegraphic Literature, Humor, Fun, Wit, and Wisdom* (New York: W. J. Johnston, 1882), 95, which identifies the author as "Miss M. J. Schofield." Josie Schofield

is identified as "our only lady operator" at the Toronto office of the Dominion Telegraph Company in the *Telegrapher,* January 16, 1875, 15.

8. "Women as Telegraph Operators," *Electrical World* 8 (June 26, 1886): 296.

9. *Telegrapher,* January 30, 1875, 27.

10. Ella Cheever Thayer, *Wired Love* (New York: W. J. Johnston, 1880). See also Gabler, *American Telegrapher,* 80, 112, 120, 128, 179.

11. "The Peculiarities of Telegraphers in the Early and Later Periods," *Telegrapher,* August 23, 1873, 205.

12. Lida A. Churchill, *My Girls* (Boston: D. Lothrop and Co., 1882).

13. *Operator,* October 14, 1882, 437.

14. *Telegrapher,* January 15, 1876, 13.

15. *New York Times,* October 11, 1884.

16. Henry James, *In the Cage* (Chicago: Herbert S. Stone, 1898).

17. Hall, "O. Delight Smith's Progressive Era," 171.

CHAPTER 6—WOMEN TELEGRAPHERS AND THE LABOR MOVEMENT

1. Gabler, *American Telegrapher,* 146–48.

2. *Telegrapher,* April 18, 1868, 274.

3. Ibid, July 15, 1876, 169.

4. Andrews, "What the Girls Can Do," 116.

5. *Telegrapher,* January 22, 1870, 173; Gabler, *American Telegrapher,* 148–49; Ulriksson, *Telegraphers,* 20–28.

6. News Clippings 1869–72, Series 7, Box 35, Folder 5, Western Union Collection; *Chicago Tribune,* January 4, 1870.

7. *Journal of the Telegraph,* May 1, 1869, 134. *Telegrapher,* September 23, 1876, 234.

8. *Telegrapher,* January 15, 1870, 168, 165.

9. Ibid., January 22, 1870, 173.

10. Ibid., February 19, 1870, 208; January 29, 1870, 184.

11. Foner, *Women and the American Labor Movement,* 114, 146; Andrews, "'What the Girls Can Do,'" 114.

12. Quoted in Diane Balser, *Sisterhood and Solidarity: Feminism and Labor in Modern Times* (Boston: South End Press, 1987), 27.

13. Garland, "Women as Telegraphists," 251–52.

14. Kieve, *Electric Telegraph,* 186–87.

15. Foner, *Women and the American Labor Movement,* 163; Gabler, *American Telegrapher,* 71–72, 159–63.

16. *Chicago Tribune,* July 15, 16, 1883.

17. Gabler, *American Telegrapher,* 6–7; Ulriksson, *Telegraphers,* 33.

18. *Chicago Tribune,* July 17, 1883.

19. Ibid., July 18, 1883.

20. Gabler, *American Telegrapher,* 9.

21. Ibid., 137–40; *Chicago Tribune,* July 20, 21, 1883.

22. *Charlotte (N.C.) Home and Democrat,* August 3, 1883; *Raleigh (N.C.) News and Observer,* August 10, 1883; *Concord (N.C.) Register,* August 17, 1883.

23. *Chicago Tribune*, July 20, 1883.

24. *New York World*, July 20, 1883.

25. Ibid., July 19, 23, 28, 1883.

26. *New York Times*, July 25, August 9, 1883; Gabler, *American Telegrapher*, 143.

27. Gabler, *American Telegrapher*, 164.

28. *Anson Times* (Wadesboro, N.C.), August 16, 1883.

29. *New York World*, August 19, 1883.

30. *New York Times*, September 7, 1883; Foner, *Women and the American Labor Movement*, 192.

31. *New York World*, August 19, 1883.

32. Hagemann, "Feminism and the Sexual Division of Labour," 143–46.

33. McIsaac, *Order of Railroad Telegraphers*, 5–19; *Railroad Telegrapher*, May 1905, 607.

34. Ulriksson, *Telegraphers*, 67–70.

35. *Chicago Tribune*, August 10, 1907.

36. Butler, *Women and the Trades*, 293–94.

37. Ulriksson, *Telegraphers*, 71; *Telegraph Age*, June 1, 1905, 235.

38. *San Francisco Chronicle*, June 20, 1907.

39. Ibid., June 21, 1907.

40. Ibid., June 22, 1907.

41. Ulriksson, *Telegraphers*, 72–77.

42. *Commercial Telegraphers' Journal*, August 1907, 810; Ulriksson, *Telegraphers*, 78; *San Francisco Chronicle*, August 8, 1907; *Washington Post*, August 10, 1907.

43. Ulriksson, *Telegraphers*, 80; *Chicago Tribune*, August 9–10, 1907.

44. *Chicago Tribune*, August 11, 1907, Foner, *Women and the American Labor Movement*, 219–20, 299, 304–5.

45. *Commercial Telegraphers' Journal*, August 1907, 847–88.

46. Ibid., September 1907, 974; *New York Times*, August 26, 1907.

47. *Chicago Tribune*, August 18, 1907.

48. Ibid., August 10, 1907.

49. Ibid., August 11, 1907.

50. *Commercial Telegraphers' Journal*, June 1907, 606; September 1907, 944.

51. Ibid., September 1907, 977.

52. *Chicago Tribune*, August 11, 1907.

53. Ibid., August 10, 11, 13, 1907.

54. Ibid., August 18, 1907.

55. Foner, *Women and the American Labor Movement*, 316.

56. Margaret Dreier Robins to Mary Dreier, September 6, 1907, Papers of the WTUL, Section 3, Reel 20.

57. Ibid., September 12, 1907.

58. *Chicago Tribune*, August 19, 1907; Ulriksson, *Telegraphers*, 83.

59. Margaret Dreier Robins to Mary Dreier, October 24, 1907, Papers of the WTUL, section 3, reel 20; *Rockford (Ill.) Daily Register-Gazette*, October 17, 1907.

60. Ulriksson, *Telegraphers*, 83.

61. *Commercial Telegraphers' Journal*, October 1907, 1061.

62. Butler, *Women and the Trades*, 294; Ulriksson, *Telegraphers*, 89.

63. Ulriksson, *Telegraphers*, 91.

64. Foner, *Women and the American Labor Movement*, 477; correspondence, Lynne Belluscio to author, January 14, 1993; Hall, "O. Delight Smith's Progressive Era," 171–74.

65. Jepsen, *Ma Kiley*, 70.

66. Ulriksson, *Telegraphers*, 111–13; *New York Times*, June 12, 1919.

67. *Commercial Telegraphers' Journal*, August 1919, 387.

68. Ibid., September 1919, 458–59; February 1920, 55.

69. Ibid., November 1919, 529.

70. Rabenseifner, "Die Frau im oeffentlichen Dienst," 227–28.

71. "Postal Telegraph Company Signs Union Agreement," *Commercial Telegraphers' Journal*, March 1937.

72. Ibid., August–September 1937, 75.

73. Gary M. Fink, ed., *The Greenwood Encyclopedia of American Institutions: Labor Unions* (Westport, Conn.: Greenwood Press, 1977), 373–75; Ulriksson, *Telegraphers*, 175–89; *New York Times*, January 8, 9, 1946.

74. Fink, ed., *Greenwood Encyclopedia of American Institutions*, 373–75; "History of Transportation-Communications International Union (TCU)," http:// members.aol.com/ tcucarmen/tcuhist.htm.

CHAPTER 7—CONCLUSIONS

1. Tillotson, "'We may all soon be "first-class men,"'" 103.

2. Bouvier, *Histoire des dames employées dans les postes, télégraphes et téléphones*, 121–26; Hagemann, "Feminism and the Sexual Division of Labour," 149.

3. Rabenseifner, "Die Frau im oeffentlichen Dienst," 226.

4. Ulriksson, *Telegraphers*, 129; Alvin F. Harlow, *Old Wires and New Waves* (New York: Appleton-Century, 1936), 506.

5. Ulriksson, *Telegraphers*, 169.

6. Carrie Glasser, "Some Problems in the Development of the Communications Industry," *American Economic Review* 35 (September 1945): 598.

7. Jepsen, *Ma Kiley*, 47.

8. Carolyn Marvin, *When Old Technologies Were New* (New York: Oxford University Press, 1989), 3.

9. U.S. Bureau of the Census, *Statistical Abstract of the United States: 1996* (Washington. D.C., 1996), 406.

10. Ruth Perry and Lisa Greber, "Women and Computers: An Introduction," *Signs* 16 (1990): 85–87.

11. William Gibson, *Neuromancer* (New York: Ace Books, 1984).

12. "Gender and Labor History: Learning from the Past, Looking to the Future," in *Work Engendered: Toward a New History of American Labor*, ed. Ava Baron (Ithaca, N.Y.: Cornell University Press, 1991), 4.

Bibliography

Adams, Ramon F. *The Language of the Railroader.* Norman: University of Oklahoma Press, 1977.

Ahvenainen, Jorma. "The Far Eastern Telegraphs: The History of Telegraphic Communications between the Far East, Europe and America before the First World War." *Suomalaisen Tiedeakatemian Toimituksia*, Sarja-Ser. B, Nide-Tom 216 (1981).

Anderson, Margo. "The History of Women and the History of Statistics." *Journal of Women's History* 4 (spring 1992): 14–36.

Andrews, Melodie. "'What the Girls Can Do': The Debate over the Employment of Women in the Early American Telegraph Industry." *Essays in Economic and Business History* 8 (1990): 109–20.

Annals of Cleveland. Cleveland: Works Progress Administration in Ohio, 1938.

Aron, Cindy Sondik. *Ladies and Gentlemen of the Civil Service: Middle-Class Workers in Victorian America.* New York: Oxford University Press, 1987.

Baark, Erik. *Lightning Wires: The Telegraph and China's Technological Modernization, 1860–1890.* Westport, Conn.: Greenwood Press, 1997.

Baker, Elizabeth Faulkner. *Technology and Women's Work.* New York: Columbia University Press, 1964.

Baron, Ava, ed. *Work Engendered: Toward a New History of American Labor.* Ithaca, N.Y.: Cornell University Press, 1991.

Bartky, Ian R. "Running on Time." *Railroad History* 159 (autumn 1988): 18–38.

Berry, Ralph Edward. *The Work of Juniors in the Telegraph Service.* Berkeley: University of California Division of Vocational Education, 1922.

Blondheim, Menahem. *News over the Wires: The Telegraph and the Flow of Public Information in America, 1844–1897.* Cambridge, Mass.: Harvard University Press, 1994.

Bouvier, Jeanne. *Histoire des dames employées dans les postes, télégraphes et téléphones de 1714 à 1929.* Paris: Presses Universitaires de France, 1930.

Brock, Gerald W. *The Telecommunications Industry: The Dynamics of Market Structure.* Cambridge, Mass.: Harvard University Press, 1981.

Brodie, Janet Farrell. *Contraception and Abortion in Nineteenth-Century America.* Ithaca, N.Y.: Cornell University Press, 1994.

Brooks, John. *Telephone: The First Hundred Years.* New York: Harper & Row, 1976.

Buckingham, Charles. "The Telegraph of To-day." *Scribner's Magazine*, July 1889, 3–22.

Burman, Shirley. "Women and the American Railroad—Documentary Photography." *Journal of the West*, April 1994, 36–41.

Burman, Shirley, curator. "Women and the American Railroad: 135 Years of Women's Association with the Railroad." Wilmington, N.C., Railroad Museum, April 2–May 31, 1995. A photographic and interpretive exhibition of women's association with the railroads.

Butler, Elizabeth Beardsley. *Women and the Trades: Pittsburgh 1907–8*. Pittsburgh: University of Pittsburgh Press, 1984.

Cammon, H. M. "Women's Work and Wages." *Harper's Magazine*, April 1869, 665–70.

Carnegie, Andrew. *Autobiography*. Boston: Houghton Mifflin, 1920.

Carter, Kate. "The Story of Telegraphy." In *Our Pioneer Heritage*, vol. 4. Salt Lake City: Daughters of Utah Pioneers, 1961.

Charlotte (N.C.) Home and Democrat.

Chicago Tribune.

Churchill, Lida A. *My Girls*. Boston: D. Lothrop and Co., 1882.

Clarke, Thomas Curtis, et al. *The American Railway: Its Construction, Management, and Appliances*. 1897; reprint, New York: Arno Press, 1976.

Coe, Lewis. *The Telegraph: A History of Morse's Invention and Its Predecessors in the United States*. Jefferson, N.C.: McFarland, 1993.

Commercial Telegraphers' Union of America Journal.

Compendium of the Tenth Census: 1880. Washington, D.C.: U.S. Government Printing Office, 1888.

Conot, Robert. *A Streak of Luck*. New York: Seaview Books, 1979.

Cranch, C. P. "An Evening with the Telegraph Wires." *Atlantic Monthly*, September 1858, 490–94.

"The Dangers of Wired Love." *Electrical World*, February 13, 1886, 68–69.

Davies, Margery. *Woman's Place Is at the Typewriter*. Philadelphia: Temple University Press, 1982.

Douglas, Paul H. *Real Wages in the United States, 1890–1926*. Boston: Houghton Mifflin, 1930.

Dye, Nancy Schrom. "The Women's Trade Union League of New York, 1903–1920." Ph.D. dissertation, University of Wisconsin, 1974; Ann Arbor, Michigan: University Microfilms International, 1984.

The Eighth Census of the United States: 1860. Washington, D.C.: U.S. Government Printing Office, 1864.

Ezra Cornell Papers. Division of Rare and Manuscript Collections, Cornell University, Ithaca, N.Y.

"The First Woman Operator." *Telegraph and Telephone Age*, October 1, 1910, 659–60.

Fison, Roger. "Is Morse Telegraphy Doomed to Extinction?" *Railroad Man's Magazine*, May 1917, 60–77.

Foner, Philip. *Women and the American Labor Movement: From Colonial Times to the Eve of World War I*. New York: Free Press, 1979.

Gabler, Edwin. *The American Telegrapher: A Social History, 1860–1900*. New Brunswick, N.J.: Rutgers University Press, 1988.

Garland, Charles. "Women as Telegraphists." *Economic Journal* 11 (June 1901): 251–61.

Gilman, Amy. "'Cogs to the Wheels': The Ideology of Women's Work in Mid-19th-Century Fiction." In *Hidden Aspects of Women's Work*, edited by Christine Bose, Roslyn Feldberg, and Natalie Sokoloff. New York: Praeger, 1987.

Gouvernement Général de L'Afrique Occidentale Française. *Les Postes et télégraphes en Afrique Occidentale.* Corbeil, France: Ed. Crete, Imprimerie Typographique, 1907.

Hagemann, Gro. "Feminism and the Sexual Division of Labour: Female Labour in the Norwegian Telegraph Service around the Turn of the Century," *Scandinavian Journal of History* 10 (1985): 143–54.

Hall, Jacquelyn Dowd. "O. Delight Smith's Progressive Era: Labor, Feminism, and Reform in the Urban South." In *Visible Women: New Essays on American Activism,* edited by Nancy A. Hewitt and Suzanne Lebsock. Champaign: University of Illinois Press, 1993.

Handbuch der Frauenarbeit in Oesterreich. Herausgegeben von der Kammer fuer Arbeiter und Angestellte in Wien. Vienna: Carl Ueberreuter, 1930.

Harlow, Alvin F. *Old Wires and New Waves.* New York: Appleton-Century, 1936.

History of Tooele County. Salt Lake City: Tooele County Daughters of Utah Pioneers, 1961.

Holcombe, Lee. *Victorian Ladies at Work: Middle-Class Working Women in England and Wales, 1850–1914.* Hamden, Conn.: Archon Books, 1973.

Israel, Paul. *From Machine Shop to Industrial Laboratory: Telegraphers and the Changing Context of American Invention, 1830–1920.* Baltimore: Johns Hopkins University Press, 1992.

James, Edward T., ed. *The Papers of the Women's Trade Union League and Its Principal Leaders.* Woodbridge, Conn.: Research Publications, 1981. Microfilm.

James, Henry. *In the Cage.* Chicago: Herbert S. Stone, 1898.

Jepsen, Thomas C. *Ma Kiley: The Life of a Railroad Telegrapher.* El Paso: Texas Western Press, 1997.

———. "The Telegraph Comes to Colorado: A New Technology and Its Consequences." *Essays and Monographs in Colorado History* no.7 (1987): 1–25.

———. "Two 'Lightning Slingers' from South Carolina." *South Carolina Historical Magazine* 94 (October 1993): 264–82.

———. "Women Telegraphers in the Railroad Depot." *Railroad History* 173 (autumn 1995): 142–54.

———. "Women Telegraph Operators on the Western Frontier." *Journal of the West* 35 (April 1996): 72–80.

Johnson, John J. "Pioneer Telegraphy in Chile, 1852–1876." *Stanford University Publications, University Series, History, Economics, and Political Science* 66, no. 1 (1948).

Josephson, Hannah. *The Golden Threads: New England's Mill Girls and Magnates.* New York: Duell, Sloan and Pearce, 1949.

Josserand, Peter. "Lap Orders." *Railroad Magazine,* September 1942, 89.

Journal of the Telegraph.

Karbelashvily, A. "Europe-India Telegraph 'Bridge' via the Caucasus." *Telecommunications Journal* 56 (1989): 719–23.

Kessler-Harris, Alice. *A Woman's Wage: Historical Meanings and Social Consequences.* Lexington: University Press of Kentucky, 1990.

Kieve, Jeffrey. *The Electric Telegraph: A Social and Economic History.* Newton Abbot, Devon: David & Charles, 1973.

Kiley, Ma. "The Bug and I." Parts 1–4. *Railroad Magazine,* April–July 1950.

Lewistown (Pa.) Sentinel.

Long Beach (Calif.) Press.

Mabee, Carleton. *American Leonardo: A Life of Samuel F. B. Morse.* New York: Knopf, 1944.

Marvin, Carolyn. *When Old Technologies Were New.* New York: Oxford University Press, 1989.

McCarthy, Justin. "Along the Wires." *Harper's New Monthly Magazine,* February 1870, 416–21.

McCullough, David. *The Johnstown Flood.* New York: Simon and Schuster, 1968.

McGaw, Judith A., ed. *Early American Technology: Making and Doing Things from the Colonial Era to 1850.* Chapel Hill: University of North Carolina Press, 1994.

McIsaac, Archibald M. *The Order of Railroad Telegraphers: A Study in Trade Unionism and Collective Bargaining.* Princeton: Princeton University Press, 1933.

Mitchell, Minnie Swan. "The Lingo of Telegraph Operators." *American Speech* 12 (April 1937): 154–55.

Moreau, Louise R. "The Feminine Touch in Telecommunications." *Antique Wireless Association Review* 4 (1989): 70–83.

Napa (Calif.) Register.

Nevada State Journal.

New York Times.

The Ninth Census of the United States: 1870. Washington, D.C.: U.S. Government Printing Office, 1872.

Norwood, Stephen H. *Labor's Flaming Youth: Telephone Operators and Worker Militancy, 1878–1923.* Urbana: University of Illinois Press, 1990.

Oates, Stephen B. *Confederate Cavalry West of the River.* Austin: University of Texas Press, 1992.

O'Connor, Richard. *Johnstown: The Day the Dam Broke.* New York: J. B. Lippincott, 1957.

"The Oldest Lady Telegrapher." *Telegraph Age,* September 16, 1897, 382.

Operator.

Our Life, 1882–1982, Akron, Iowa. Akron, Iowa: *Akron Register-Tribune* and *Le Mars Daily Sentinel* Job Printing, 1982.

Penny, Virginia. *How Women Can Make Money.* Springfield, Mass.: Fisk, 1870.

Perry, Ruth, and Lisa Greber. "Women and Computers: An Introduction." *Signs* 16 (autumn 1990): 85–87.

Phillips, Barnet. "The Thorsdale Telegraphs." *Atlantic Monthly,* October 1876, 400–417.

Plum, William R. *The Military Telegraph during the Civil War in the United States.* 2 vols. Chicago: Jansen, McClurg, 1882.

Prescott, George B. *History, Theory, and Practice of the Electric Telegraph.* Boston: Ticknor and Fields, 1860.

Rabenseifner, Anna. "Die Frau im oeffentlichen Dienst." *Handbuch der Frauenarbeit in Oesterreich.* Vienna: Carl Ueberreuter, 1930.

Railroad Retirement Board. Record of Employee's Prior Service for Mattie C. Kuhn, August 1941.

Railroad Telegrapher.

Rayne, Martha L. *What Can a Woman Do? or, Her Position in the Business and Literary World.* Petersburgh, N.Y.: Eagle, 1893.

Reid, James D. *The Telegraph in America: Its Founders, Promoters, and Noted Men.* New York: Derby Brothers, 1879.

Report of the Michigan State Commission of Inquiry into Wages and the Conditions of Labor for Women and the Advisability of Establishing a Minimum Wage. Lansing, Mich.: Wynkoop Hallenbeck Crawford, 1915.

Rosenthal, Eric, ed. *Encyclopedia of Southern Africa.* London: Frederick Warne, 1973.

Ruiz, Ramon Eduardo. *Triumphs and Tragedy: A History of the Mexican People.* New York: Norton, 1992.

Rules and Instructions for the Information and Guidance of the Employees of the Western Union Telegraph Company. New York: Russell Brothers, 1870.

San Francisco Chronicle.

Schofield, Josie. "Wooing by Wire." *Telegrapher,* November 20, 1875, 277–78; November 27, 1875, 283–84.

Selden, Bernice. *The Mill Girls.* New York: Atheneum, 1983.

Shanks, William F. G. "Women's Work and Wages." *Harper's Magazine,* September 1868, 546–53.

Shridharani, Krishnalal. *Story of the Indian Telegraphs: A Century of Progress.* New Delhi: Government of India Press, 1953.

Shiers, George, ed. *The Electric Telegraph: An Historical Anthology.* New York: Arno Press, 1977.

Statistical Abstract of the United States: 1996. Washington, D.C.: U.S. Bureau of the Census, 1996.

"The Telegraph." *Harper's Magazine,* August 1873, 332–60.

Telegraph Age (Telegraph and Telephone Age after 1909)

Telegrapher.

Thayer, Ella Cheever. *Wired Love.* New York: W. J. Johnston, 1880.

Thompson, Robert L. *Wiring a Continent: The History of the Telegraph Industry in the United States, 1832–1866.* Princeton: Princeton University Press, 1947.

Tillotson, Shirley. "The Operators along the Coast: A Case Study of the Link between Gender, Skilled Labour and Social Power, 1900–1930." *Acadiensis* 20 (1990): 72–88.

———. "'We may all soon be "first-class men"': Gender and Skill in Canada's Early Twentieth Century Urban Telegraph Industry." *Labour/Le Travail* 27 (Spring 1991): 97–125.

Towers, Walter Kellogg. *From Beacon Fire to Radio: The Story of Long-Distance Communication.* New York: Harper & Brothers, 1924.

Turbin, Carole. "Reconceptualizing Family, Work, and Labor Organizing: Working Women in Troy, 1860–1890." In *Hidden Aspects of Women's Work,* ed. Christine Bose, Roslyn Feldberg, and Natalie Sokoloff. New York: Praeger, 1987.

Ulriksson, Vidkunn. *The Telegraphers: Their Craft and Their Unions.* Washington, D.C.: Public Affairs Press, 1953.

Weingarten, Ruthe. *Texas Women: A Pictorial History from Indians to Astronauts.* Austin: Eakin Press, 1985.

West Chester (Pa.) Daily Local News.

Western Union Telegraph Company Collection, 1848–1963. Archives Center, National Museum of American History, Smithsonian Institution, Washington, D.C.

Withuhn, William L., ed. *Rails across America: A History of Railroads in North America.*
    New York: Smithmark, 1993.
Wolff, Michael F. "The Marriage That Almost Was." *IEEE Spectrum* 13 (February 1976):
    40–51.
"Women as Telegraph Operators." *Electrical World*, June 26, 1886, 296.
Wright, Carroll D. *The Working Girls of Boston.* Boston: Wright and Potter, 1889.
Wright, Helena. "Sarah G. Bagley: A Biographical Note." *Labor History* 20 (summer
    1979): 398–413.

Index